# Eureka Math
# Grade 4
# Modules 6 & 7

Special thanks go to the Gordon A. Cain Center and to the Department of Mathematics at Louisiana State University for their support in the development of *Eureka Math*.

**Published by the non-profit Great Minds**

Copyright © 2015 Great Minds. No part of this work may be reproduced, sold, or commercialized, in whole or in part, without written permission from Great Minds. Non-commercial use is licensed pursuant to a Creative Commons Attribution-NonCommercial-ShareAlike 4.0 license; for more information, go to http://greatminds.net/maps/math/copyright. "Great Minds" and "Eureka Math" are registered trademarks of Great Minds.

Printed in the U.S.A.

This book may be purchased from the publisher at eureka-math.org

10 9 8 7 6 5 4 3

ISBN 978-1-63255-306-5

Name _____   Date _____

1. Shade the first 7 units of the tape diagram. Count by tenths to label the number line using a fraction and a decimal for each point. Circle the decimal that represents the shaded part.

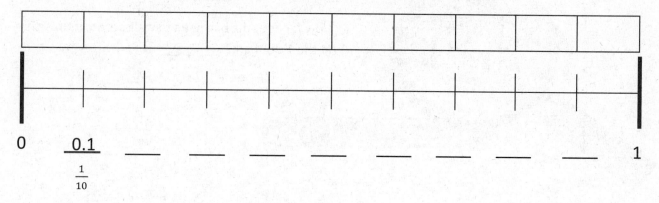

0        0.1        ___   ___   ___   ___   ___   ___   ___   ___   1

         $\frac{1}{10}$

2. Write the total amount of water in fraction form and decimal form. Shade the last bottle to show the correct amount.

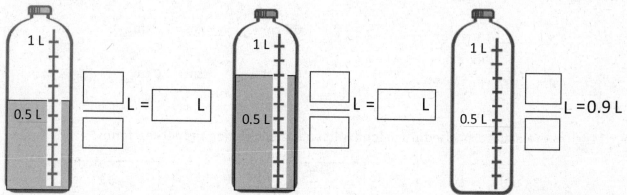

$\frac{\phantom{0}}{\phantom{0}}$ L = [       ] L        $\frac{\phantom{0}}{\phantom{0}}$ L = [       ] L        $\frac{\phantom{0}}{\phantom{0}}$ L = 0.9 L

3. Write the total weight of the food on each scale in fraction form or decimal form.

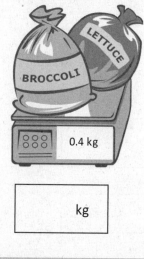   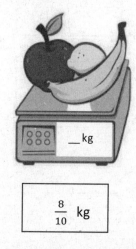   

0.4 kg        ___ kg        0.5 kg

[       ] kg        [  $\frac{8}{10}$  ] kg        [       ] kg

Lesson 1:   Use metric measurement to model the decomposition of one whole into tenths.

©2015 Great Minds. eureka-math.org
G4-M6M7-SE-B4-1.3.1-1.2016

4.  Write the length of the bug in centimeters.  (The drawing is not to scale.)

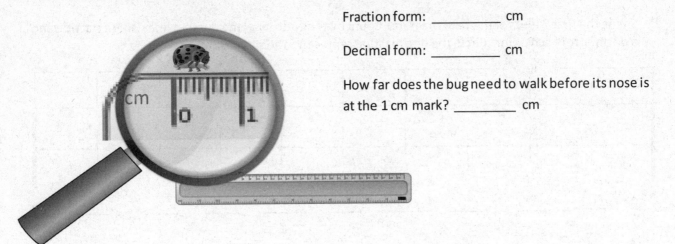

Fraction form: _____ cm

Decimal form: _____ cm

How far does the bug need to walk before its nose is at the 1 cm mark? _____ cm

5.  Fill in the blank to make the sentence true in both fraction form and decimal form.

a.  $\frac{8}{10}$ cm + _____ cm = 1 cm                 0.8 cm + _____ cm = 1.0 cm

b.  $\frac{2}{10}$ cm + _____ cm = 1 cm                 0.2 cm + _____ cm = 1.0 cm

c.  $\frac{6}{10}$ cm + _____ cm = 1 cm                 0.6 cm + _____ cm = 1.0 cm

6.  Match each amount expressed in unit form to its equivalent fraction and decimal forms.

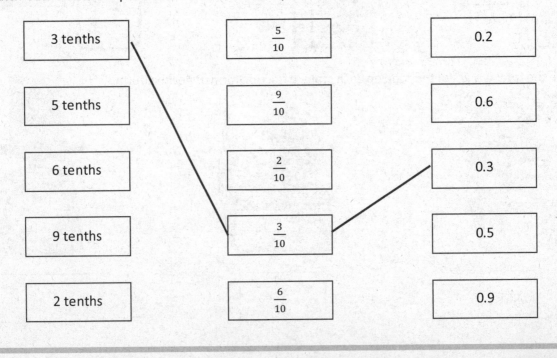

| 3 tenths | | $\frac{5}{10}$ | | 0.2 |
| 5 tenths | | $\frac{9}{10}$ | | 0.6 |
| 6 tenths | | $\frac{2}{10}$ | | 0.3 |
| 9 tenths | | $\frac{3}{10}$ | | 0.5 |
| 2 tenths | | $\frac{6}{10}$ | | 0.9 |

**Lesson 1:**    Use metric measurement to model the decomposition of one whole into tenths.

EUREKA MATH

©2015 Great Minds. eureka-math.org
G4-M6M7-SE-B4-1.3.1-1.2016

Name _____ Date _____

Shade the first 4 units of the tape diagram. Count by tenths to label the number line using a fraction and a decimal for each point. Circle the decimal that represents the shaded part.

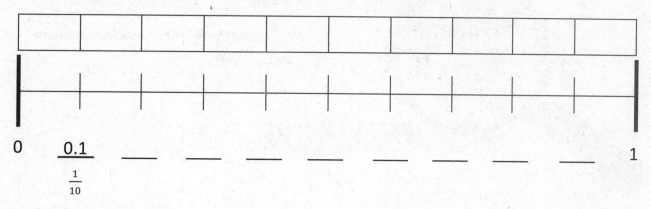

0        0.1        ___   ___   ___   ___   ___   ___   ___   ___        1

         $\frac{1}{10}$

2.  Write the total amount of water in fraction form and decimal form. Shade the last bottle to show the correct amount.

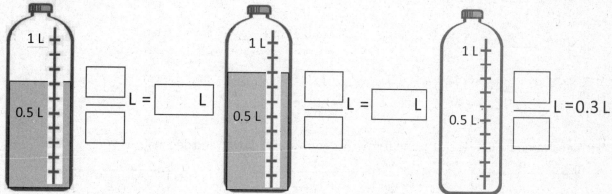

3.  Write the total weight of the food on each scale in fraction form or decimal form.

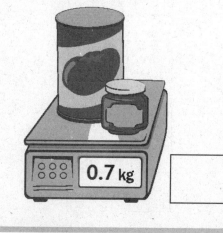

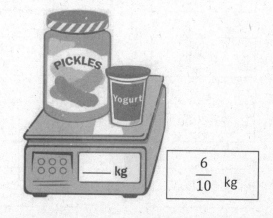

EUREKA
MATH

Lesson 1:     Use metric measurement to model the decomposition of one whole into tenths.

3

©2015 Great Minds. eureka-math.org
G4-M6M7-SE-B4-1.3.1-1.2016

4.  Write the length of the bug in centimeters.  (The drawing is not to scale.)

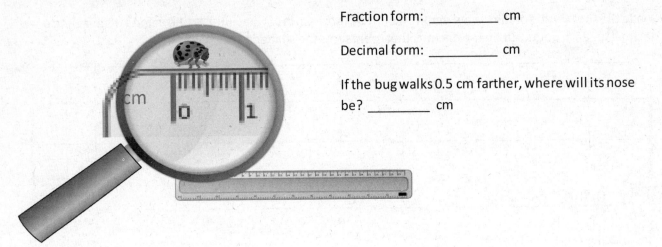

Fraction form: _____ cm

Decimal form: _____ cm

If the bug walks 0.5 cm farther, where will its nose be? _____ cm

5.  Fill in the blank to make the sentence true in both fraction and decimal form.

a.  $\frac{4}{10}$ cm + _____ cm = 1 cm

0.4 cm + _____ cm = 1.0 cm

b.  $\frac{3}{10}$ cm + _____ cm = 1 cm

0.3 cm + _____ cm = 1.0 cm

c.  $\frac{8}{10}$ cm + _____ cm = 1 cm

0.8 cm + _____ cm = 1.0 cm

6.  Match each amount expressed in unit form to its equivalent fraction and decimal.

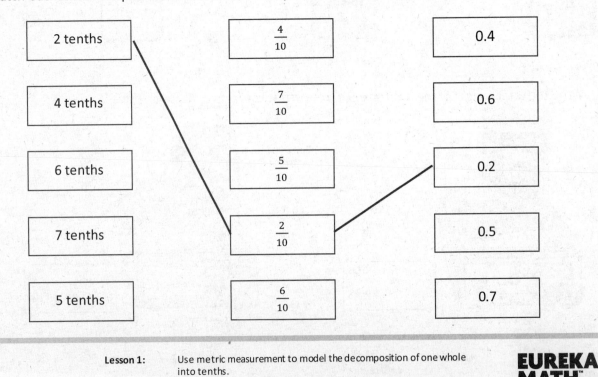

| 2 tenths | | $\frac{4}{10}$ | | 0.4 |
| 4 tenths | | $\frac{7}{10}$ | | 0.6 |
| 6 tenths | | $\frac{5}{10}$ | | 0.2 |
| 7 tenths | | $\frac{2}{10}$ | | 0.5 |
| 5 tenths | | $\frac{6}{10}$ | | 0.7 |

Lesson 1:   Use metric measurement to model the decomposition of one whole into tenths.

 EUREKA MATH

Name _____    Date _____

1. For each length given below, draw a line segment to match. Express each measurement as an equivalent mixed number.

   a.  2.6 cm

   b.  3.4 cm

   c.  3.7 cm

   d.  4.2 cm

   e.  2.5 cm

2. Write the following as equivalent decimals. Then, model and rename the number as shown below.

   a.  2 ones and 6 tenths = _____

   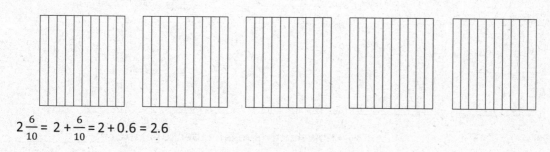

   $2\frac{6}{10} = 2 + \frac{6}{10} = 2 + 0.6 = 2.6$

EUREKA
MATH™

Lesson 2:    Use metric measurement and area models to represent tenths as
             fractions greater than 1 and decimal numbers.

5

©2015 Great Minds. eureka-math.org
G4-M6M7-SE-B4-1.3.1-1.2016

b.   4 ones and 2 tenths = _____

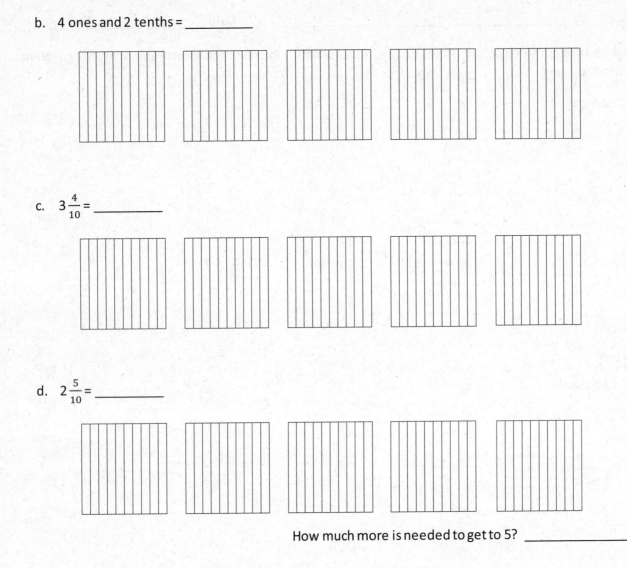

c.   $3\frac{4}{10}$ = _____

d.   $2\frac{5}{10}$ = _____

How much more is needed to get to 5? _____

e.   $\frac{37}{10}$ = _____

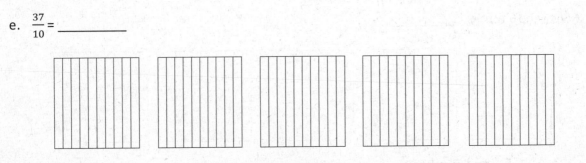

How much more is needed to get to 5? _____

Lesson 2:     Use metric measurement and area models to represent tenths as
fractions greater than 1 and decimal numbers.

EUREKA MATH

©2015 Great Minds. eureka-math.org
G4-M6M7-SE-B4-1.3.1-1.2016

Name _____    Date _____

1.  For each length given below, draw a line segment to match.  Express each measurement as an equivalent mixed number.

    a.   2.6 cm

    b.   3.5 cm

    c.   1.7 cm

    d.   4.3 cm

    e.   2.2 cm

2.  Write the following in decimal form.  Then, model and rename the number as shown below.

    a.   2 ones and 4 tenths = _____

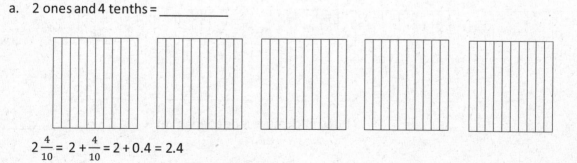

$$2\frac{4}{10} = 2 + \frac{4}{10} = 2 + 0.4 = 2.4$$

Lesson 2:    Use metric measurement and area models to represent tenths as
            fractions greater than 1 and decimal numbers.

7

©2015 Great Minds. eureka-math.org
G4-M6M7-SE-B4-1.3.1-1.2016

b.  3 ones and 8 tenths = _____

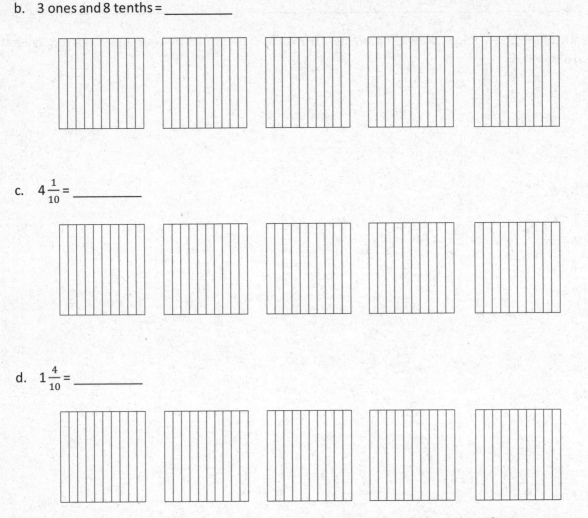

c.  $4\frac{1}{10}$ = _____

d.  $1\frac{4}{10}$ = _____

How much more is needed to get to 5? _____

e.  $\frac{33}{10}$ = _____

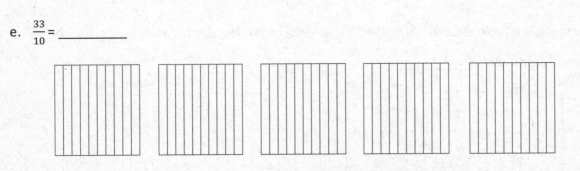

How much more is needed to get to 5? _____

**Lesson 2:**     Use metric measurement and area models to represent tenths as
fractions greater than 1 and decimal numbers.

tenths area model

Lesson 2:   Use metric measurement and area models to represent tenths as
fractions greater than 1 and decimal numbers.

9

This page intentionally left blank

Name _____   Date _____

1.  Circle groups of tenths to make as many ones as possible.

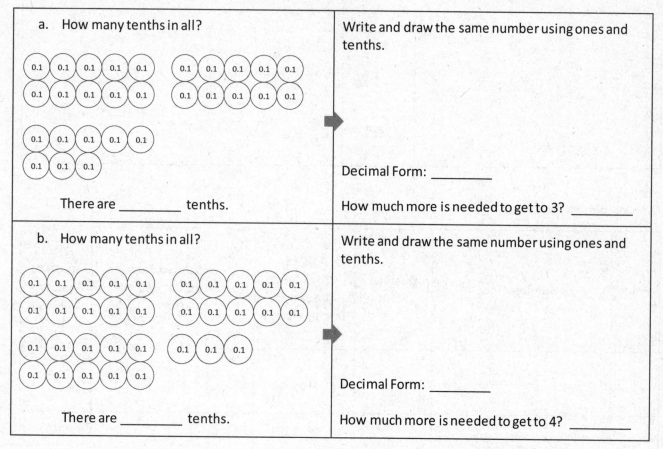

a.  How many tenths in all?

There are _____ tenths.

Write and draw the same number using ones and tenths.

Decimal Form: _____

How much more is needed to get to 3? _____

b.  How many tenths in all?

There are _____ tenths.

Write and draw the same number using ones and tenths.

Decimal Form: _____

How much more is needed to get to 4? _____

2.  Draw disks to represent each number using tens, ones, and tenths.  Then, show the expanded form of the number in fraction form and decimal form as shown.  The first one has been completed for you.

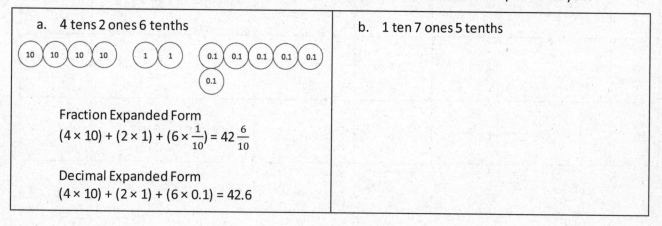

a.  4 tens 2 ones 6 tenths

Fraction Expanded Form

$(4 \times 10) + (2 \times 1) + (6 \times \frac{1}{10}) = 42\frac{6}{10}$

Decimal Expanded Form

$(4 \times 10) + (2 \times 1) + (6 \times 0.1) = 42.6$

b.  1 ten 7 ones 5 tenths

Lesson 3:      Represent mixed numbers with units of tens, ones, and tenths with place value disks, on the number line, and in expanded form.

| | |
|---|---|
| c.   2 tens 3 ones 2 tenths | d.   7 tens 4 ones 7 tenths |

3.  Complete the chart.

| Point | Number Line | Decimal Form | Mixed Number (ones and fraction form) | Expanded Form (fraction or decimal form) | How much to get to the next one? |
|---|---|---|---|---|---|
| a. | | | $3\frac{9}{10}$ | | 0.1 |
| b. | | | | | |
| c. | | | | $(7 \times 10) + (4 \times 1) + (7 \times \frac{1}{10})$ | |
| d. | | | $22\frac{2}{10}$ | | |
| e. | | | | $(8 \times 10) + (8 \times 0.1)$ | |

©2015 Great Minds. eureka-math.org
G4-M6M7-SE-B4-1.3.1-1.2016

Name _____    Date _____

1.  Circle groups of tenths to make as many ones as possible.

    a.  How many tenths in all?

    0.1  0.1  0.1  0.1  0.1    0.1  0.1  0.1  0.1
    0.1  0.1  0.1  0.1  0.1

    There are _____ tenths.

    Write and draw the same number using ones and tenths.

    Decimal Form: _____

    How much more is needed to get to 2? _____

    b.  How many tenths in all?

    0.1  0.1  0.1  0.1  0.1    0.1  0.1  0.1  0.1  0.1
    0.1  0.1  0.1  0.1  0.1    0.1  0.1  0.1  0.1  0.1
    0.1  0.1  0.1  0.1  0.1

    There are _____ tenths.

    Write and draw the same number using ones and tenths.

    Decimal Form: _____

    How much more is needed to get to 3? _____

2.  Draw disks to represent each number using tens, ones, and tenths. Then, show the expanded form of the number in fraction form and decimal form as shown. The first one has been completed for you.

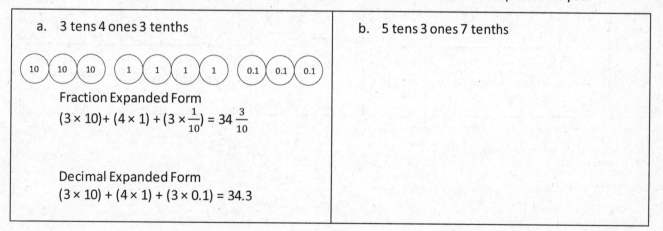

    a.  3 tens 4 ones 3 tenths

    10  10  10   1  1  1  1   0.1  0.1  0.1

    Fraction Expanded Form
    $(3 \times 10) + (4 \times 1) + (3 \times \frac{1}{10}) = 34\frac{3}{10}$

    Decimal Expanded Form
    $(3 \times 10) + (4 \times 1) + (3 \times 0.1) = 34.3$

    b.  5 tens 3 ones 7 tenths

EUREKA MATH™

Lesson 3:    Represent mixed numbers with units of tens, ones, and tenths with place value disks, on the number line, and in expanded form.

13

| c. 3 tens 2 ones 3 tenths | d. 8 tens 4 ones 8 tenths |
|---|---|
| | |

3. Complete the chart.

| Point | Number Line | Decimal Form | Mixed Number (ones and fraction form) | Expanded Form (fraction or decimal form) | How much to get to the next one? |
|---|---|---|---|---|---|
| a. | | | $4\frac{6}{10}$ | | |
| b. | (point between 24 and 25) | | | | 0.5 |
| c. | | | | $(6 \times 10) + (3 \times 1) + (6 \times \frac{1}{10})$ | |
| d. | | | $71\frac{3}{10}$ | | |
| e. | | | | $(9 \times 10) + (9 \times 0.1)$ | |

Lesson 3:   Represent mixed numbers with units of tens, ones, and tenths with place value disks, on the number line, and in expanded form.

©2015 Great Minds. eureka-math.org
G4-M6M7-SE-B4-1.3.1-1.2016

| Point | Number Line | Decimal Form | Mixed Number (ones and fraction form) | Expanded Form (fraction or decimal form) | How much more is needed to get to the next one? |
|-------|-------------|--------------|----------------------------------------|-------------------------------------------|---------------------------------------------------|
| a. | | | | | |
| b. | | | | | |
| c. | | | | | |
| d. | | | | | |

tenths on a number line

Lesson 3: Represent mixed numbers with units of tens, ones, and tenths with place value disks, on the number line, and in expanded form.

15

©2015 Great Minds. eureka-math.org
G4-M6M7-SE-B4-1.3.1-1.2016

This page intentionally left blank

Name _____     Date _____

1.  a.  What is the length of the shaded part of the meter stick in centimeters?

                                                                    1 meter

    b.  What fraction of a meter is 1 centimeter?

    c.  In fraction form, express the length of the shaded portion of the meter stick.

                                                                    1 meter

    d.  In decimal form, express the length of the shaded portion of the meter stick.

    e.  What fraction of a meter is 10 centimeters?

2.  Fill in the blanks.

    a.  1 tenth = _____ hundredths      b.  $\frac{1}{10}$ m = $\frac{}{100}$ m      c.  $\frac{2}{10}$ m = $\frac{20}{}$ m

3.  Use the model to add the shaded parts as shown. Write a number bond with the total written in decimal form and the parts written as fractions. The first one has been done for you.

    a.                                          1 meter

                                                                    0.13
                                                                   ╱    ╲
                                                                  $\frac{1}{10}$   $\frac{3}{100}$

        $\frac{1}{10}$ m + $\frac{3}{100}$ m = $\frac{13}{100}$ m = 0.13 m

b.

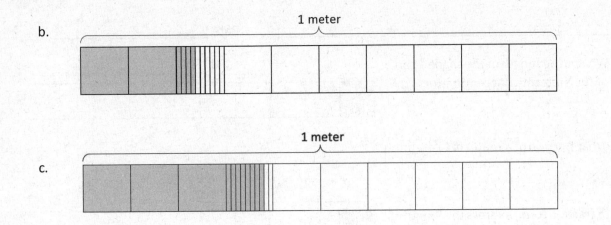

c.

4.  On each meter stick, shade in the amount shown.  Then, write the equivalent decimal.

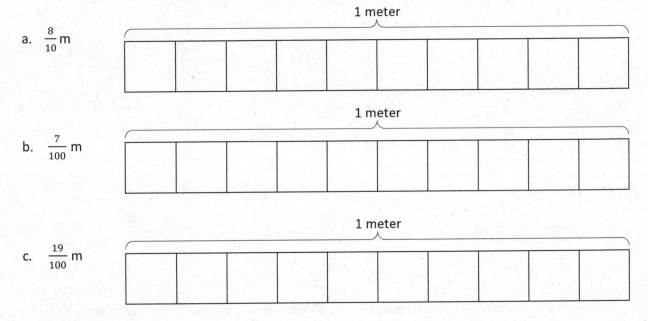

a.  $\frac{8}{10}$ m

b.  $\frac{7}{100}$ m

c.  $\frac{19}{100}$ m

5.  Draw a number bond, pulling out the tenths from the hundredths as in Problem 3.  Write the total as the equivalent decimal.

a.  $\frac{19}{100}$ m

b.  $\frac{28}{100}$ m

c.  $\frac{77}{100}$

d.  $\frac{94}{100}$

Lesson 4:    Use meters to model the decomposition of one whole into hundredths.
            Represent and count hundredths.

Name _____    Date _____

1.  a.  What is the length of the shaded part of the meter stick in centimeters?

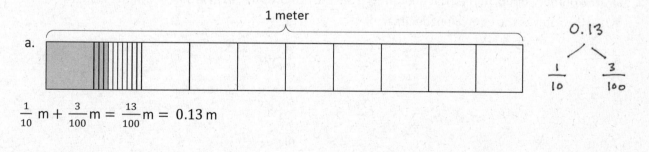

1 meter

    b.  What fraction of a meter is 3 centimeters?

    c.  In fraction form, express the length of the shaded portion of the meter stick.

1 meter

    d.  In decimal form, express the length of the shaded portion of the meter stick.

    e.  What fraction of a meter is 30 centimeters?

2.  Fill in the blanks.

    a.  5 tenths = _____ hundredths

    b.  $\dfrac{5}{10}$ m = $\dfrac{}{100}$ m

    c.  $\dfrac{4}{10}$ m = $\dfrac{40}{}$ m

3.  Use the model to add the shaded parts as shown.  Write a number bond with the total written in decimal form and the parts written as fractions.  The first one has been done for you.

1 meter

    a.

0.13

$\dfrac{1}{10}$    $\dfrac{3}{100}$

$\dfrac{1}{10}$ m + $\dfrac{3}{100}$ m = $\dfrac{13}{100}$ m = 0.13 m

EUREKA
MATH™

Lesson 4:    Use meters to model the decomposition of one whole into hundredths.
              Represent and count hundredths.

19

b.

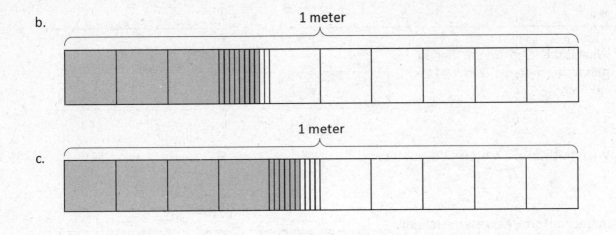

c.

4. On each meter stick, shade in the amount shown. Then, write the equivalent decimal.

a. $\frac{9}{10}$ m

b. $\frac{15}{100}$ m

c. $\frac{41}{100}$ m

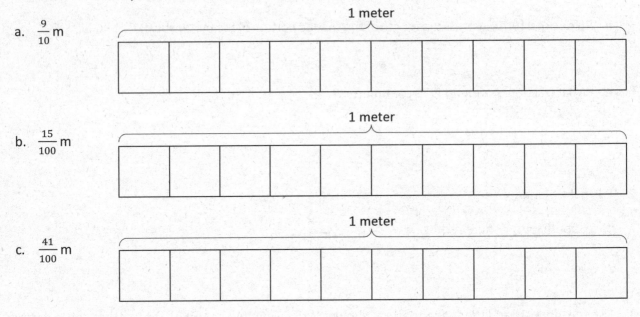

5. Draw a number bond, pulling out the tenths from the hundredths, as in Problem 3 of the Homework. Write the total as the equivalent decimal.

a. $\frac{23}{100}$ m

b. $\frac{38}{100}$ m

c. $\frac{82}{100}$

d. $\frac{76}{100}$

Lesson 4:    Use meters to model the decomposition of one whole into hundredths. Represent and count hundredths.

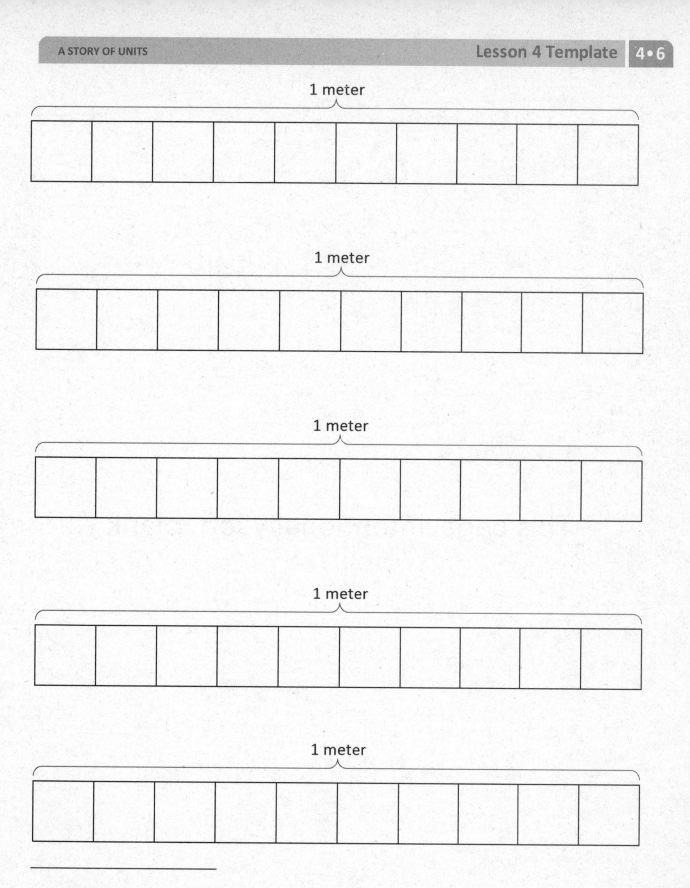

1 meter

1 meter

1 meter

1 meter

1 meter

tape diagram in tenths

**Lesson 4:** Use meters to model the decomposition of one whole into hundredths. Represent and count hundredths.

©2015 Great Minds. eureka-math.org
G4-M6M7-SE-B4-1.3.1-1.2016

This page intentionally left blank

Name _____ Date _____

1. Find the equivalent fraction using multiplication or division. Shade the area models to show the equivalency. Record it as a decimal.

a. $\dfrac{3 \times}{10 \times} = \dfrac{}{100}$
b. $\dfrac{50 \div}{100 \div} = \dfrac{}{10}$

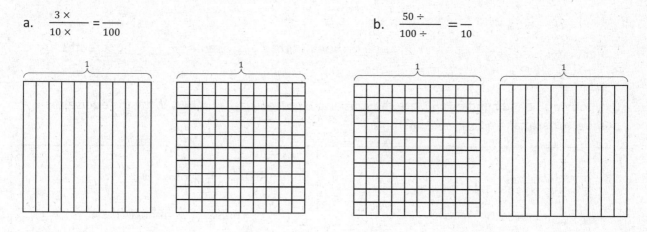

2. Complete the number sentences. Shade the equivalent amount on the area model, drawing horizontal lines to make hundredths.

a. 37 hundredths = _____ tenths + _____ hundredths

   Fraction form: _____

   Decimal form: _____

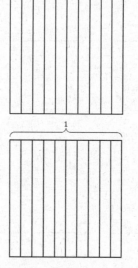

b. 75 hundredths = _____ tenths + _____ hundredths

   Fraction form: _____

   Decimal form: _____

3. Circle hundredths to compose as many tenths as you can. Complete the number sentences. Represent each with a number bond as shown.

a.

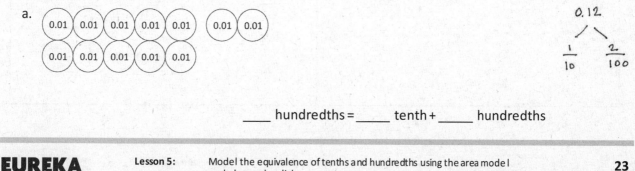

_____ hundredths = _____ tenth + _____ hundredths

EUREKA MATH

Lesson 5: Model the equivalence of tenths and hundredths using the area model and place value disks.

23

b.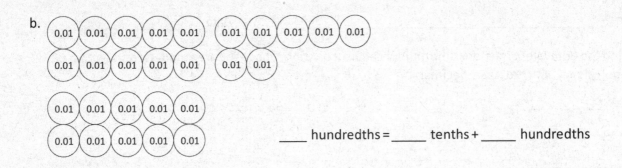

_____ hundredths = _____ tenths + _____ hundredths

4. Use both tenths and hundredths place value disks to represent each number. Write the equivalent number in decimal, fraction, and unit form.

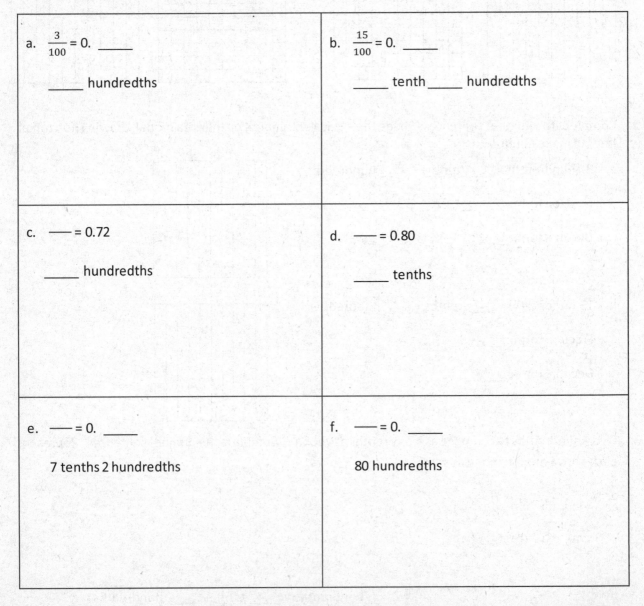

| a. $\frac{3}{100}$ = 0. _____ <br><br> _____ hundredths | b. $\frac{15}{100}$ = 0. _____ <br><br> _____ tenth _____ hundredths |
|---|---|
| c. ――― = 0.72 <br><br> _____ hundredths | d. ――― = 0.80 <br><br> _____ tenths |
| e. ――― = 0. _____ <br><br> 7 tenths 2 hundredths | f. ――― = 0. _____ <br><br> 80 hundredths |

**Lesson 5:** Model the equivalence of tenths and hundredths using the area model and place value disks.

EUREKA MATH™

Name _____     Date _____

1. Find the equivalent fraction using multiplication or division. Shade the area models to show the equivalency. Record it as a decimal.

   a. $\dfrac{4 \times}{10 \times} = \dfrac{}{100}$

   b. $\dfrac{60 \div}{100 \div} = \dfrac{}{10}$

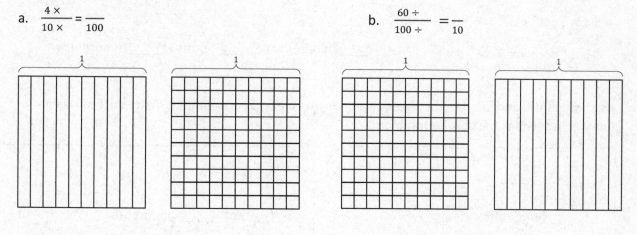

2. Complete the number sentences. Shade the equivalent amount on the area model, drawing horizontal lines to make hundredths.

   a. 36 hundredths = _____ tenths + _____ hundredths

      Decimal form: _____

      Fraction form: _____

   b. 82 hundredths = _____ tenths + _____ hundredths

      Decimal form: _____

      Fraction form: _____

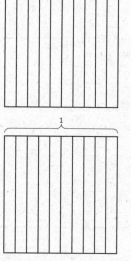

3. Circle hundredths to compose as many tenths as you can. Complete the number sentences. Represent each with a number bond as shown.

   a.

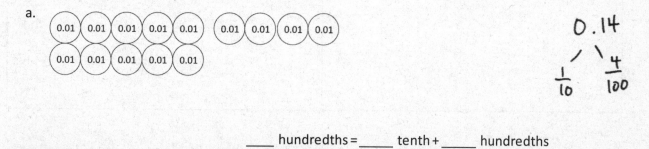

   _____ hundredths = _____ tenth + _____ hundredths

Lesson 5:  Model the equivalence of tenths and hundredths using the area model and place value disks.

25

©2015 Great Minds. eureka-math.org
G4-M6M7-SE-B4-1.3.1-1.2016

b.

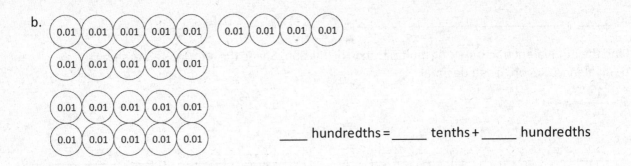

\_\_\_\_ hundredths = \_\_\_\_ tenths + \_\_\_\_ hundredths

4. Use both tenths and hundredths place value disks to represent each number. Write the equivalent number in decimal, fraction, and unit form.

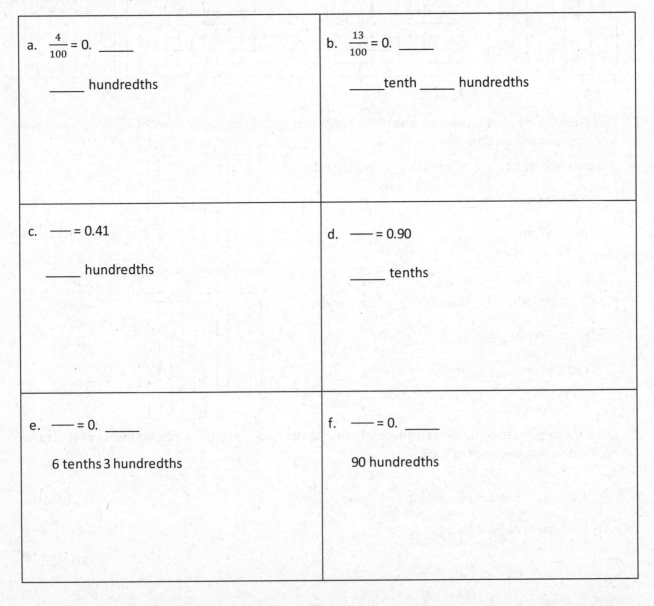

| | |
|---|---|
| a. $\frac{4}{100} = 0.$ \_\_\_\_ <br><br> \_\_\_\_ hundredths | b. $\frac{13}{100} = 0.$ \_\_\_\_ <br><br> \_\_\_\_ tenth \_\_\_\_ hundredths |
| c. —— $= 0.41$ <br><br> \_\_\_\_ hundredths | d. —— $= 0.90$ <br><br> \_\_\_\_ tenths |
| e. —— $= 0.$ \_\_\_\_ <br><br> 6 tenths 3 hundredths | f. —— $= 0.$ \_\_\_\_ <br><br> 90 hundredths |

  **Lesson 5:**  Model the equivalence of tenths and hundredths using the area model and place value disks.

EUREKA MATH™

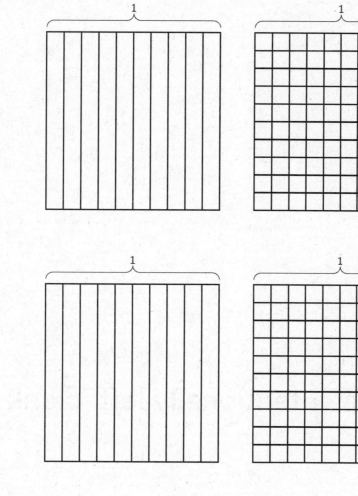

tenths and hundredths area model

Lesson 5:    Model the equivalence of tenths and hundredths using the area model
            and place value disks.

27

©2015 Great Minds. eureka-math.org
G4-M6M7-SE-B4-1.3.1-1.2016

This page intentionally left blank

Name _____   Date _____

1.  Shade the area models to represent the number, drawing horizontal lines to make hundredths as needed. Locate the corresponding point on the number line.  Label with a point, and record the mixed number as a decimal.

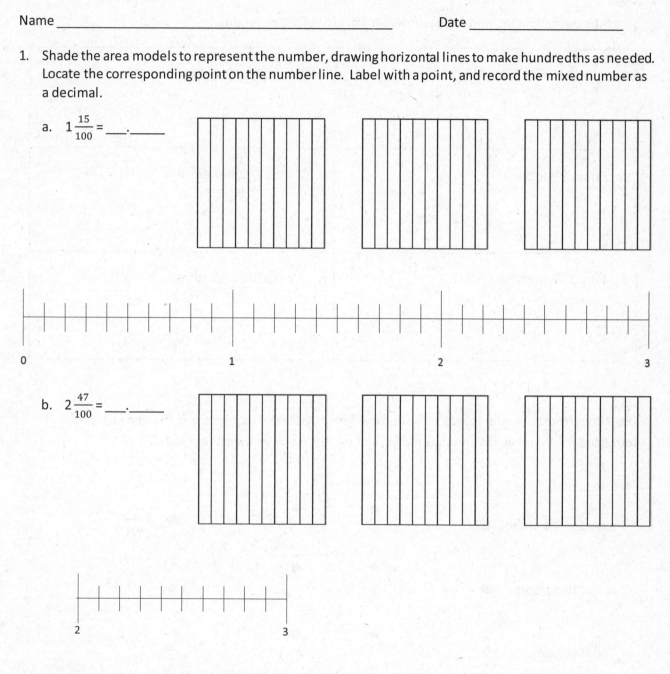

a.   $1\frac{15}{100}$ = ___.____

b.   $2\frac{47}{100}$ = ___.____

2.  Estimate to locate the points on the number lines.

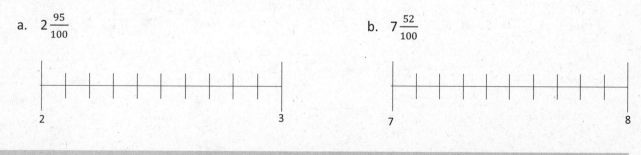

a.   $2\frac{95}{100}$

b.   $7\frac{52}{100}$

Lesson 6:   Use the area model and number line to represent mixed numbers with units of ones, tenths, and hundredths in fraction and decimal forms.

29

3.  Write the equivalent fraction and decimal for each of the following numbers.

| | |
|---|---|
| a.  1 one 2 hundredths | b.  1 one 17 hundredths |
| c.  2 ones 8 hundredths | d.  2 ones 27 hundredths |
| e.  4 ones 58 hundredths | f.  7 ones 70 hundredths |

4.  Draw lines from dot to dot to match the decimal form to both the unit form and fraction form.  All unit forms and fractions have at least one match, and some have more than one match.

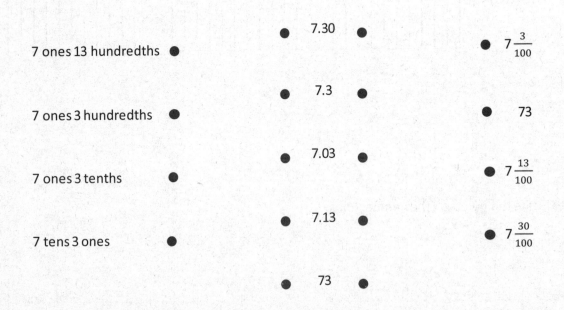

Lesson 6:     Use the area model and number line to represent mixed numbers with units of ones, tenths, and hundredths in fraction and decimal forms.

©2015 Great Minds. eureka-math.org
G4-M6M7-SE-B4-1.3.1-1.2016

Name _____     Date _____

1.  Shade the area models to represent the number, drawing horizontal lines to make hundredths as needed. Locate the corresponding point on the number line.  Label with a point, and record the mixed number as a decimal.

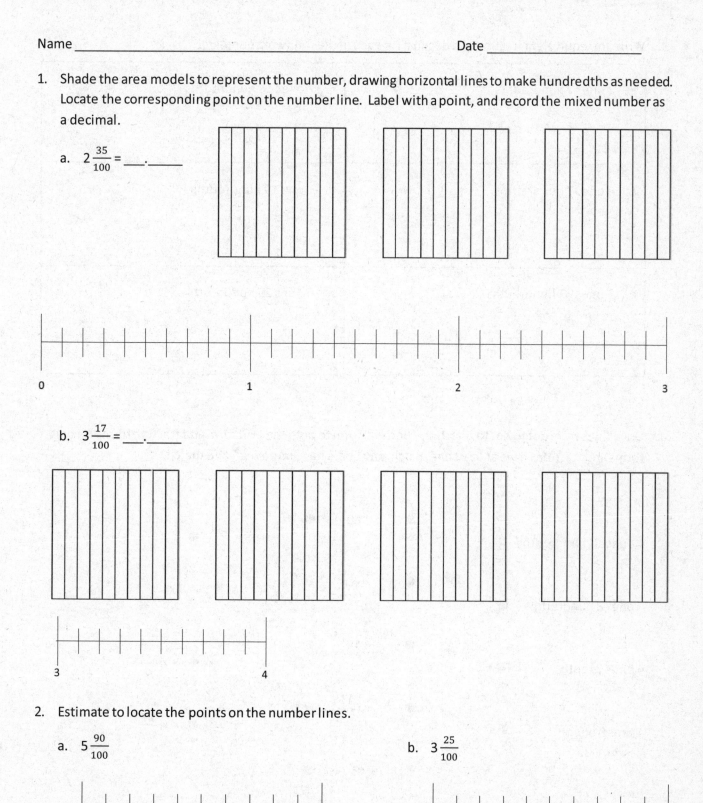

a.   $2\frac{35}{100} =$ ___._____

b.   $3\frac{17}{100} =$ ___._____

2.  Estimate to locate the points on the number lines.

a.   $5\frac{90}{100}$

b.   $3\frac{25}{100}$

EUREKA
MATH™

Lesson 6:  Use the area model and number line to represent mixed numbers with
units of ones, tenths, and hundredths in fraction and decimal forms.

31

©2015 Great Minds. eureka-math.org
G4-M6M7-SE-B4-1.3.1-1.2016

3. Write the equivalent fraction and decimal for each of the following numbers.

| | |
|---|---|
| a.  2 ones 2 hundredths | b.  2 ones 16 hundredths |
| c.  3 ones 7 hundredths | d.  1 one 18 hundredths |
| e.  9 ones 62 hundredths | f.  6 ones 20 hundredths |

4. Draw lines from dot to dot to match the decimal form to both the unit form and fraction form.  All unit forms and fractions have at least one match, and some have more than one match.

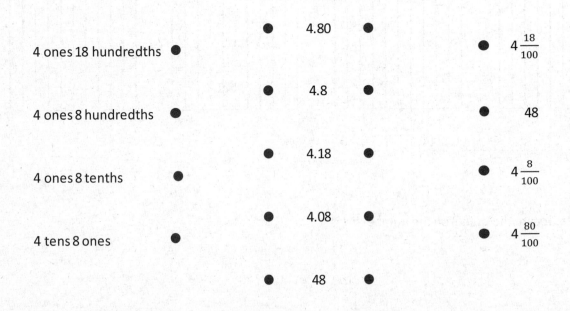

Lesson 6:     Use the area model and number line to represent mixed numbers with units of ones, tenths, and hundredths in fraction and decimal forms.

©2015 Great Minds. eureka-math.org
G4-M6M7-SE-B4-1.3.1-1.2016

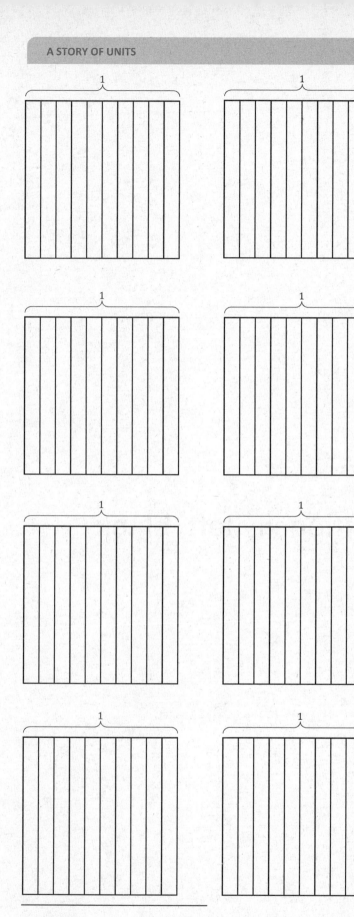

area model

**Lesson 6:** Use the area model and number line to represent mixed numbers with units of ones, tenths, and hundredths in fraction and decimal forms.

33

©2015 Great Minds. eureka-math.org
G4-M6M7-SE-B4-1.3.1-1.2016

This page  intentionally left  blank

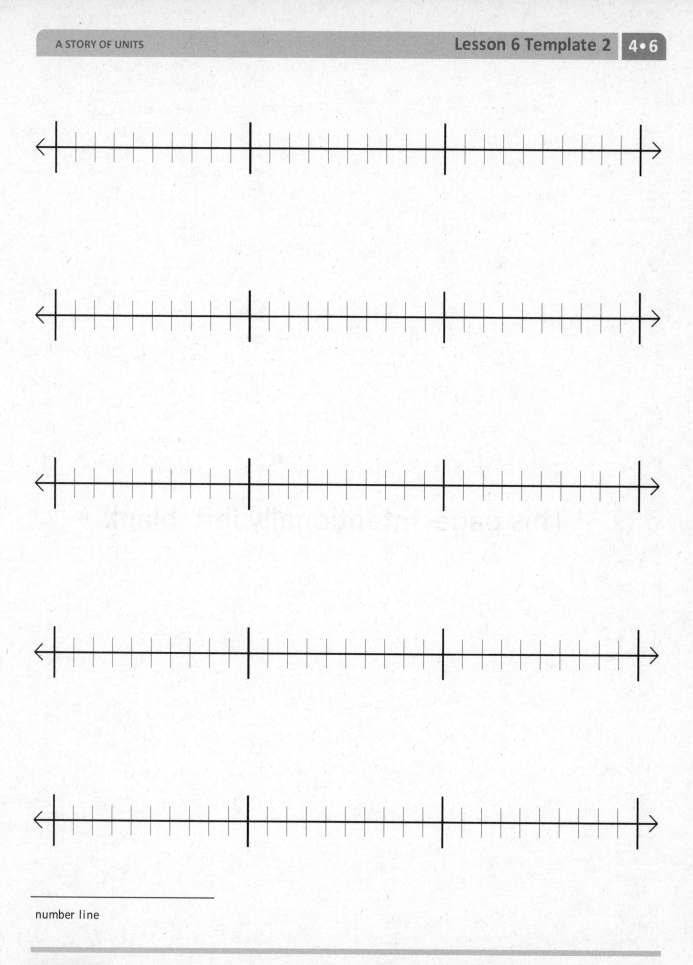

number line

EUREKA MATH

Lesson 6:    Use the area model and number line to represent mixed numbers with
            units of ones, tenths, and hundredths in fraction and decimal forms.

©2015 Great Minds. eureka-math.org
G4-M6M7-SE-B4-1.3.1-1.2016

This page intentionally left blank

Name _____    Date _____

1. Write a decimal number sentence to identify the total value of the place value disks.

   a.

   | 10 | 10 | | 0.1 | 0.1 | 0.1 | 0.1 | 0.1 | | 0.01 | 0.01 | 0.01 |

        2 tens         5 tenths        3 hundredths

   _____ + _____ + _____ = _____

   b.

   | 100 | 100 | 100 | 100 | 100 | | 0.01 | 0.01 | 0.01 | 0.01 |

        5 hundreds        4 hundredths

   _____ + _____ = _____

2. Use the place value chart to answer the following questions. Express the value of the digit in unit form.

| hundreds | tens | ones | . | tenths | hundredths |
|----------|------|------|---|--------|------------|
| 4 | 1 | 6 | | 8 | 3 |

   a. The digit _____ is in the hundreds place. It has a value of _____.

   b. The digit _____ is in the tens place. It has a value of _____.

   c. The digit _____ is in the tenths place. It has a value of _____.

   d. The digit _____ is in the hundredths place. It has a value of _____.

| hundreds | tens | ones | . | tenths | hundredths |
|----------|------|------|---|--------|------------|
| 5 | 3 | 2 | | 1 | 6 |

   e. The digit _____ is in the hundreds place. It has a value of _____.

   f. The digit _____ is in the tens place. It has a value of _____.

   g. The digit _____ is in the tenths place. It has a value of _____.

   h. The digit _____ is in the hundredths place. It has a value of _____.

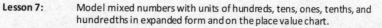

EUREKA MATH™

Lesson 7:    Model mixed numbers with units of hundreds, tens, ones, tenths, and hundredths in expanded form and on the place value chart.

37

©2015 Great Minds. eureka-math.org
G4-M6M7-SE-B4-1.3.1-1.2016

3. Write each decimal as an equivalent fraction. Then, write each number in expanded form, using both decimal and fraction notation. The first one has been done for you.

| Decimal and Fraction Form | Expanded Form | |
|---|---|---|
| | Fraction Notation | Decimal Notation |
| $15.43 = 15\frac{43}{100}$ | $(1 \times 10) + (5 \times 1) + (4 \times \frac{1}{10}) + (3 \times \frac{1}{100})$ <br><br> $10 \quad + \quad 5 \quad + \quad \frac{4}{10} \quad + \quad \frac{3}{100}$ | $(1 \times 10) + (5 \times 1) + (4 \times 0.1) + (3 \times 0.01)$ <br><br> $10 \quad + \quad 5 \quad + \quad 0.4 \quad + \quad 0.03$ |
| $21.4 = $ _____ | | |
| $38.09 = $ _____ | | |
| $50.2 = $ _____ | | |
| $301.07 = $ _____ | | |
| $620.80 = $ _____ | | |
| $800.08 = $ _____ | | |

Lesson 7:    Model mixed numbers with units of hundreds, tens, ones, tenths, and hundredths in expanded form and on the place value chart.

©2015 Great Minds. eureka-math.org
G4-M6M7-SE-B4-1.3.1-1.2016

Name _____  Date _____

1. Write a decimal number sentence to identify the total value of the place value disks.

a.

( 10 )( 10 )( 10 )    ( 0.1 )( 0.1 )( 0.1 )( 0.1 )    ( 0.01 )( 0.01 )

    3 tens            4 tenths         2 hundredths

_____ + _____ + _____ = _____

b.

( 100 )( 100 )( 100 )( 100 )    ( 0.01 )( 0.01 )( 0.01 )

     4 hundreds         3 hundredths

_____ + _____ = _____

2. Use the place value chart to answer the following questions. Express the value of the digit in unit form.

| hundreds | tens | ones | . | tenths | hundredths |
|---|---|---|---|---|---|
| 8 | 2 | 7 | | 6 | 4 |

a. The digit _____ is in the hundreds place. It has a value of _____.

b. The digit _____ is in the tens place. It has a value of _____.

c. The digit _____ is in the tenths place. It has a value of _____.

d. The digit _____ is in the hundredths place. It has a value of _____.

| hundreds | tens | ones | . | tenths | hundredths |
|---|---|---|---|---|---|
| 3 | 4 | 5 | | 1 | 9 |

e. The digit _____ is in the hundreds place. It has a value of _____.

f. The digit _____ is in the tens place. It has a value of _____.

g. The digit _____ is in the tenths place. It has a value of _____.

h. The digit _____ is in the hundredths place. It has a value of _____.

EUREKA MATH™

Lesson 7:   Model mixed numbers with units of hundreds, tens, ones, tenths, and hundredths in expanded form and on the place value chart.

39

©2015 Great Minds. eureka-math.org
G4-M6M7-SE-B4-1.3.1-1.2016

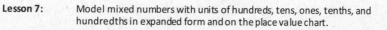

3.  Write each decimal as an equivalent fraction. Then, write each number in expanded form, using both decimal and fraction notation.  The first one has been done for you.

| Decimal and Fraction Form | Expanded Form | |
| --- | --- | --- |
| | Fraction Notation | Decimal Notation |
| $14.23 = 14\frac{23}{100}$ | $(1 \times 10) + (4 \times 1) + (2 \times \frac{1}{10}) + (3 \times \frac{1}{100})$ <br><br> $10 \ + \ 4 \ + \ \frac{2}{10} \ + \ \frac{3}{100}$ | $(1 \times 10) + (4 \times 1) + (2 \times 0.1) + (3 \times 0.01)$ <br><br> $10 \ + \ 4 \ + \ 0.2 \ + \ 0.03$ |
| 25.3 = _____ | | |
| 39.07 = _____ | | |
| 40.6 = _____ | | |
| 208.90 = _____ | | |
| 510.07 = _____ | | |
| 900.09 = _____ | | |

Lesson 7:    Model mixed numbers with units of hundreds, tens, ones, tenths, and hundredths in expanded form and on the place value chart.

| hundredths | |
| :---: | :--- |
| tenths | |
| . | |
| ones | |
| tens | |
| hundreds | |

place value chart

**Lesson 7:** Model mixed numbers with units of hundreds, tens, ones, tenths, and hundredths in expanded form and on the place value chart.

41

©2015 Great Minds. eureka-math.org
G4-M6M7-SE-B4-1.3.1-1.2016

This page  intentionally left  blank

Name _____    Date _____

1. Use the area model to represent $\frac{250}{100}$. Complete the number sentence.

   a. $\frac{250}{100}$ = _____ tenths = _____ ones _____ tenths = __.____

   b. In the space below, explain how you determined your answer to part (a).

2. Draw place value disks to represent the following decompositions:

   2 ones = _____ tenths

   | ones | . | tenths | hundredths |
   |------|---|--------|------------|
   |      |   |        |            |

   2 tenths = _____ hundredths

   | ones | . | tenths | hundredths |
   |------|---|--------|------------|
   |      |   |        |            |

   1 one 3 tenths        = ____ tenths

   | ones | . | tenths | hundredths |
   |------|---|--------|------------|
   |      |   |        |            |

   2 tenths 3 hundredths = ____ hundredths

   | ones | . | tenths | hundredths |
   |------|---|--------|------------|
   |      |   |        |            |

Lesson 8:    Use understanding of fraction equivalence to investigate decimal
             numbers on the place value chart expressed in different units.

43

©2015 Great Minds. eureka-math.org
G4-M6M7-SE-B4-1.3.1-1.2016

3. Decompose the units to represent each number as tenths.

   a. 1 = _____ tenths

   b. 2 = _____ tenths

   c. 1.7 = _____ tenths

   d. 2.9 = _____ tenths

   e. 10.7 = _____ tenths

   f. 20.9 = _____ tenths

4. Decompose the units to represent each number as hundredths.

   a. 1 = _____ hundredths

   b. 2 = _____ hundredths

   c. 1.7 = _____ hundredths

   d. 2.9 = _____ hundredths

   e. 10.7 = _____ hundredths

   f. 20.9 = _____ hundredths

5. Complete the chart. The first one has been done for you.

| Decimal | Mixed Number | Tenths | Hundredths |
|---|---|---|---|
| 2.1 | $2\frac{1}{10}$ | 21 tenths $\frac{21}{10}$ | 210 hundredths $\frac{210}{100}$ |
| 4.2 | | | |
| 8.4 | | | |
| 10.2 | | | |
| 75.5 | | | |

**Lesson 8:** Use understanding of fraction equivalence to investigate decimal numbers on the place value chart expressed in different units.

EUREKA MATH

©2015 Great Minds. eureka-math.org
G4-M6M7-SE-B4-1.3.1-1.2016

Name _____    Date _____

1. Use the area model to represent $\frac{220}{100}$. Complete the number sentence.

   a. $\frac{220}{100}$ = _____ tenths = _____ ones _____ tenths = __.____

   b. In the space below, explain how you determined your answer to part (a).

2. Draw place value disks to represent the following decompositions:

   3 ones = _____ tenths            3 tenths = _____ hundredths

   | ones | . | tenths | hundredths |
   |------|---|--------|------------|
   |      |   |        |            |

   | ones | . | tenths | hundredths |
   |------|---|--------|------------|
   |      |   |        |            |

   2 ones 3 tenths = ____ tenths       3 tenths 3 hundredths = ____ hundredths

   | ones | . | tenths | hundredths |
   |------|---|--------|------------|
   |      |   |        |            |

   | ones | . | tenths | hundredths |
   |------|---|--------|------------|
   |      |   |        |            |

Lesson 8:    Use understanding of fraction equivalence to investigate decimal numbers on the place value chart expressed in different units.

45

3. Decompose the units to represent each number as tenths.

   a.  1 = _____ tenths

   b.  2 = _____ tenths

   c.  1.3 = _____ tenths

   d.  2.6 = _____ tenths

   e.  10.3 = _____ tenths

   f.  20.6 = _____ tenths

4. Decompose the units to represent each number as hundredths.

   a.  1 = _____ hundredths

   b.  2 = _____ hundredths

   c.  1.3 = _____ hundredths

   d.  2.6 = _____ hundredths

   e.  10.3 = _____ hundredths

   f.  20.6 = _____ hundredths

5. Complete the chart. The first one has been done for you.

| Decimal | Mixed Number | Tenths | Hundredths |
|---|---|---|---|
| 4.1 | $4\frac{1}{10}$ | 41 tenths $\frac{41}{10}$ | 410 hundredths $\frac{410}{100}$ |
| 5.3 | | | |
| 9.7 | | | |
| 10.9 | | | |
| 68.5 | | | |

Lesson 8:  Use understanding of fraction equivalence to investigate decimal numbers on the place value chart expressed in different units.

©2015 Great Minds. eureka-math.org
G4-M6M7-SE-B4-1.3.1-1.2016

EUREKA MATH

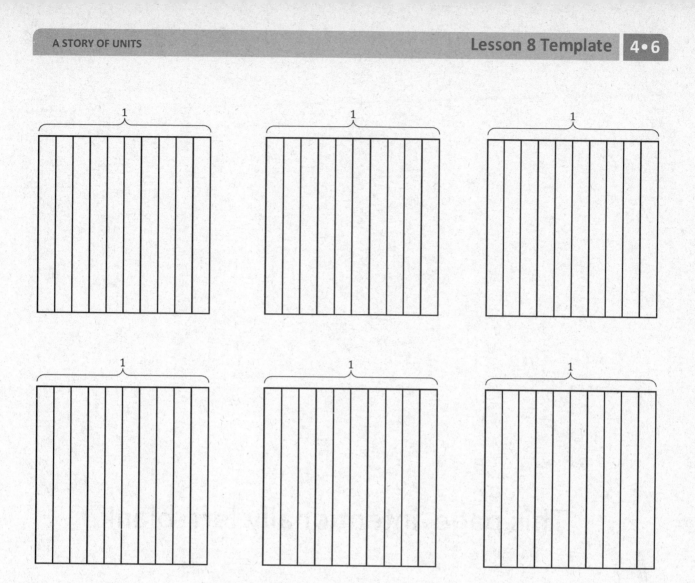

| Tens | Ones | . | Tenths | Hundredths |
|------|------|---|--------|------------|
|      |      |   |        |            |
|      |      |   |        |            |
|      |      |   |        |            |

area model and place value chart

Lesson 8:  Use understanding of fraction equivalence to investigate decimal numbers on the place value chart expressed in different units.

47

This page  intentionally left  blank

Name _____     Date _____

1. Express the lengths of the shaded parts in decimal form. Write a sentence that compares the two lengths. Use the expression *shorter than* or *longer than* in your sentence.

a.
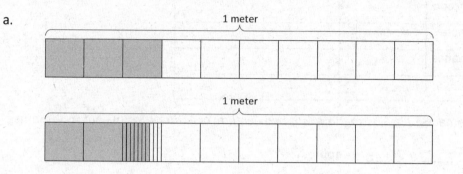
1 meter

1 meter

b.
1 meter

1 meter

c. List all four lengths from least to greatest.

2. a. Examine the mass of each item as shown below on the 1-kilogram scales. Put an X over the items that are heavier than the avocado.

    0.2 kg         0.12 kg         0.6 kg         0.61 kg

**EUREKA MATH**

**Lesson 9:**    Use the place value chart and metric measurement to compare decimals and answer comparison questions.

49

©2015 Great Minds. eureka-math.org
G4-M6M7-SE-B4-1.3.1-1.2016

b.  Express the mass of each item on the place value chart.

**Mass of Fruit (kilograms)**

| Fruit | ones | . | tenths | hundredths |
|-------|------|---|--------|------------|
| avocado | | | | |
| apple | | | | |
| bananas | | | | |
| grapes | | | | |

c.  Complete the statements below using the words *heavier than* or *lighter than* in your statements.

The avocado is _____ the apple.

The bunch of bananas is _____ the bunch of grapes.

3.  Record the volume of water in each graduated cylinder on the place value chart below.

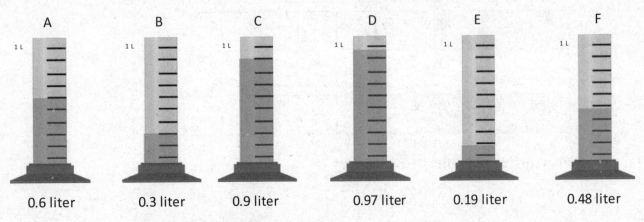

|  A | B | C | D | E | F |
|----|---|---|---|---|---|
| 0.6 liter | 0.3 liter | 0.9 liter | 0.97 liter | 0.19 liter | 0.48 liter |

**Volume of Water (liters)**

| Cylinder | ones | . | tenths | hundredths |
|----------|------|---|--------|------------|
| A | | | | |
| B | | | | |
| C | | | | |
| D | | | | |
| E | | | | |
| F | | | | |

Compare the values using >, <, or =.

a.  0.9 L _____ 0.6 L

b.  0.48 L _____ 0.6 L

c.  0.3 L _____ 0.19 L

d.  Write the volume of water in each graduated cylinder in order from least to greatest.

Lesson 9:    Use the place value chart and metric measurement to compare decimals and answer comparison questions.

Name _____ Date _____

1. Express the lengths of the shaded parts in decimal form.  Write a sentence that compares the two lengths.  Use the expression *shorter than* or *longer than* in your sentence.

    a.
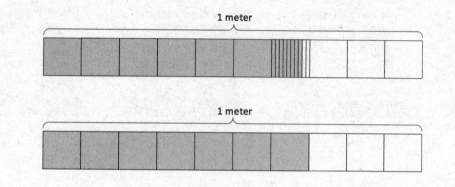

    b.

    1 meter

    1 meter

    c. List all four lengths from least to greatest.

Lesson 9:    Use the place value chart and metric measurement to compare decimals and answer comparison questions.

51

©2015 Great Minds. eureka-math.org
G4-M6M7-SE-B4-1.3.1-1.2016

2.  a.  Examine the mass of each item as shown below on the 1-kilogram scales. Put an X over the items
        that are heavier than the volleyball

       0.15 kg             0.62 kg             0.43 kg            0.25 kg

b.  Express the mass of each item on the place value chart.

### Mass of Sport Balls (kilograms)

| Sport Balls | ones | . | tenths | hundredths |
|---|---|---|---|---|
| baseball | | | | |
| volleyball | | | | |
| basketball | | | | |
| soccer ball | | | | |

c.  Complete the statements below using the words *heavier than* or *lighter than* in your statements.

The soccer ball is _____ the baseball.

The volleyball is _____ the basketball.

Lesson 9:    Use the place value chart and metric measurement to compare
decimals and answer comparison questions.

EUREKA
MATH

3. Record the volume of water in each graduated cylinder on the place value chart below.

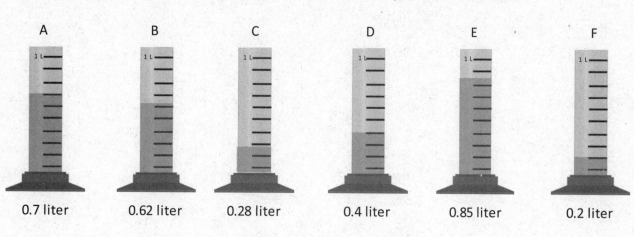

| A | B | C | D | E | F |
|---|---|---|---|---|---|
| 0.7 liter | 0.62 liter | 0.28 liter | 0.4 liter | 0.85 liter | 0.2 liter |

**Volume of Water (liters)**

| Cylinder | ones | . | tenths | hundredths |
|----------|------|---|--------|------------|
| A | | | | |
| B | | | | |
| C | | | | |
| D | | | | |
| E | | | | |
| F | | | | |

Compare the values using >, <, or =.

a.    0.4 L _____ 0.2 L

b.    0.62 L _____ 0.7 L

c.    0.2 L _____ 0.28 L

d.    Write the volume of water in each graduated cylinder in order from least to greatest.

**EUREKA MATH**

**Lesson 9:** Use the place value chart and metric measurement to compare decimals and answer comparison questions.

53

©2015 Great Minds. eureka-math.org
G4-M6M7-SE-B4-1.3.1-1.2016

This page intentionally left blank

**Mass of Rice Bags (kilograms)**

| Rice Bag | ones | . | tenths | hundredths |
|---|---|---|---|---|
| A | | | | |
| B | | | | |
| C | | | | |
| D | | | | |

**Volume of Liquid (liters)**

| Cylinder | ones | . | tenths | hundredths |
|---|---|---|---|---|
| A | | | | |
| B | | | | |
| C | | | | |
| D | | | | |

measurement record

Lesson 9: Use the place value chart and metric measurement to compare decimals and answer comparison questions.

55

This page intentionally left blank

Name _____     Date _____

1.  Shade the area models below, decomposing tenths as needed, to represent the pairs of decimal numbers. Fill in the blank with <, >, or = to compare the decimal numbers.

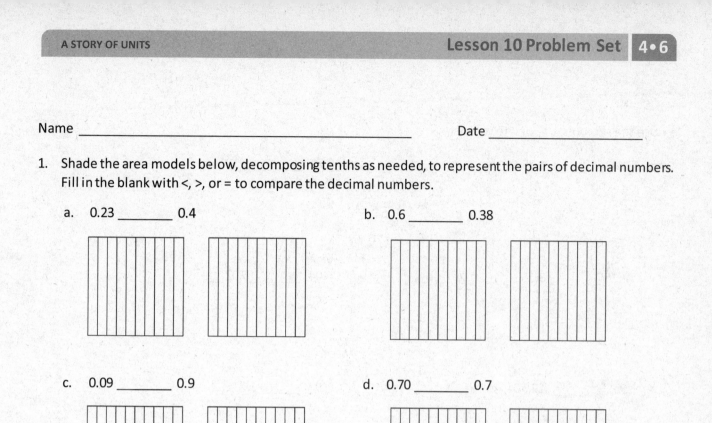

a.  0.23 _____ 0.4                          b.  0.6 _____ 0.38

c.  0.09 _____ 0.9                          d.  0.70 _____ 0.7

2.  Locate and label the points for each of the decimal numbers on the number line. Fill in the blank with <, >, or = to compare the decimal numbers.

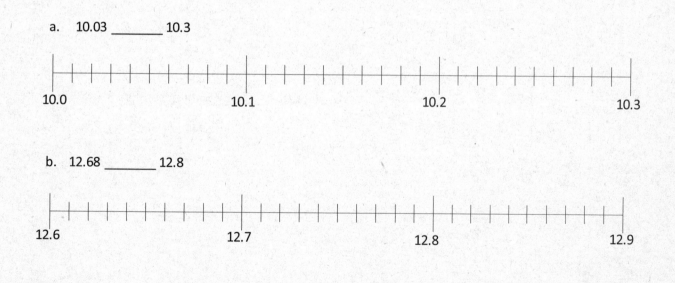

a.  10.03 _____ 10.3

b.  12.68 _____ 12.8

EUREKA MATH

Lesson 10:    Use area models and the number line to compare decimal numbers, and record comparisons using <, >, and =.

57

3.  Use the symbols <, >, or = to compare.

    a.  3.42 _____ 3.75            b.  4.21 _____ 4.12

    c.  2.15 _____ 3.15            d.  4.04 _____ 6.02

    e.  12.7 _____ 12.70           f.  1.9 _____ 1.21

4.  Use the symbols <, >, or = to compare.  Use pictures as needed to solve.

    a.  23 tenths _____ 2.3        b.  1.04 _____ 1 one and 4 tenths

    c.  6.07 _____ $6\frac{7}{10}$   d.  0.45 _____ $\frac{45}{10}$

    e.  $\frac{127}{100}$ _____ 1.72   f.  6 tenths _____ 66 hundredths

Lesson 10:    Use area models and the number line to compare decimal numbers,
              and record comparisons using <, >, and =.

©2015 Great Minds. eureka-math.org
G4-M6M7-SE-B4-1.3.1-1.2016

Name _____     Date _____

1.  Shade the parts of the area models below, decomposing tenths as needed, to represent the pairs of decimal numbers.  Fill in the blank with <, >, or = to compare the decimal numbers.

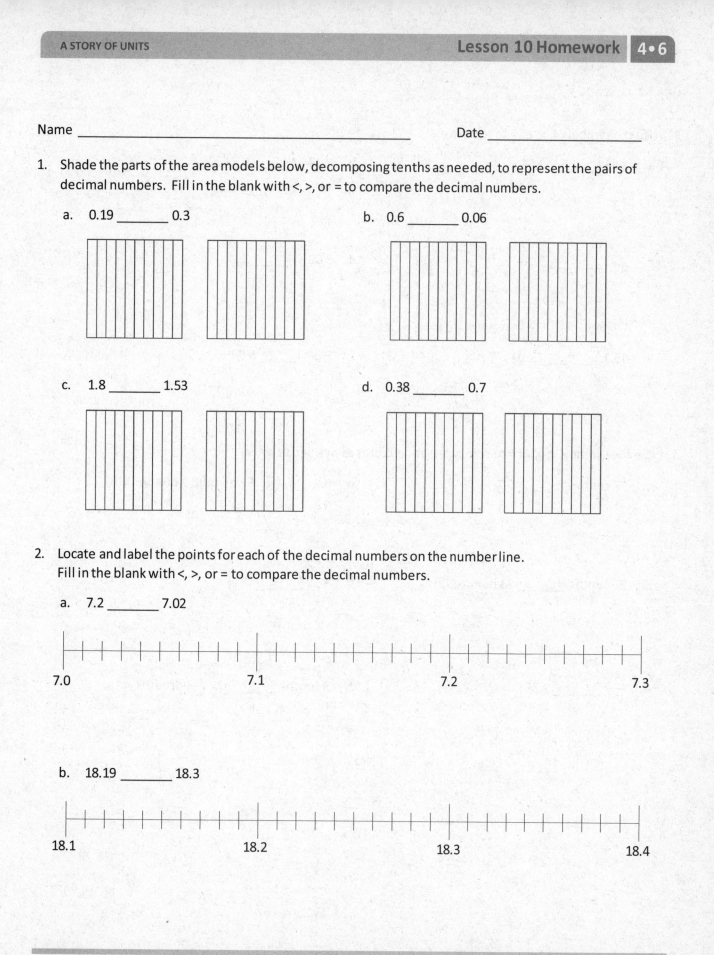

a.   0.19 _____ 0.3

b.   0.6 _____ 0.06

c.   1.8 _____ 1.53

d.   0.38 _____ 0.7

2.  Locate and label the points for each of the decimal numbers on the number line.
    Fill in the blank with <, >, or = to compare the decimal numbers.

a.   7.2 _____ 7.02

7.0          7.1               7.2          7.3

b.   18.19 _____ 18.3

18.1          18.2               18.3          18.4

EUREKA
MATH™

Lesson 10:     Use area models and the number line to compare decimal numbers,
                    and record comparisons using <, >, and =.

59

©2015 Great Minds. eureka-math.org
G4-M6M7-SE-B4-1.3.1-1.2016

3.  Use the symbols <, >, or = to compare.

    a.  2.68 _____ 2.54

    b.  6.37 _____ 6.73

    c.  9.28 _____ 7.28

    d.  3.02 _____ 3.2

    e.  13.1 _____ 13.10

    f.  5.8 _____ 5.92

4.  Use the symbols <, >, or = to compare.  Use pictures as needed to solve.

    a.  57 tenths _____ 5.7

    b.  6.2 _____ 6 ones and 2 hundredths

    c.  33 tenths _____ 33 hundredths

    d.  8.39 _____ $8\frac{39}{10}$

    e.  $\frac{236}{100}$ _____ 2.36

    f.  3 tenths _____ 22 hundredths

Lesson 10:    Use area models and the number line to compare decimal numbers, and record comparisons using <, >, and =.

EUREKA MATH™

comparing with area models

**Lesson 10:**   Use area models and the number line to compare decimal numbers,
and record comparisons using <, >, and =.

61

©2015 Great Minds. eureka-math.org
G4-M6M7-SE-B4-1.3.1-1.2016

This page  intentionally left  blank

Name _____ Date _____

1. Plot the following points on the number line.

   a. 0.2, $\frac{1}{10}$, 0.33, $\frac{12}{100}$, 0.21, $\frac{32}{100}$

   b. 3.62, 3.7, $3\frac{85}{100}$, $\frac{38}{10}$, $\frac{364}{100}$

   c. $6\frac{3}{10}$, 6.31, $\frac{628}{100}$, $\frac{62}{10}$, 6.43, 6.40

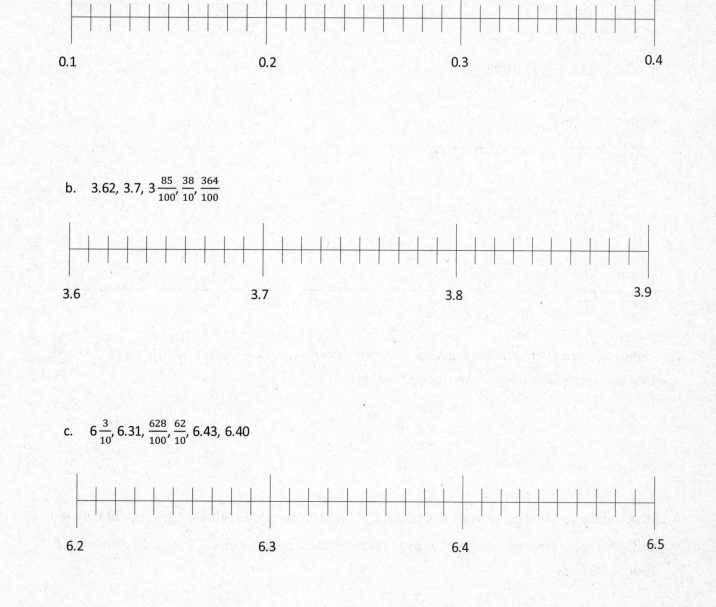

EUREKA MATH

Lesson 11: Compare and order mixed numbers in various forms.

63

2. Arrange the following numbers in order from greatest to least using decimal form. Use the > symbol between each number.

a. $\frac{27}{10}$, 2.07, $\frac{27}{100}$, $2\frac{71}{100}$, $\frac{227}{100}$, 2.72

b. $12\frac{3}{10}$, 13.2, $\frac{134}{100}$, 13.02, $12\frac{20}{100}$

c. $7\frac{34}{100}$, $7\frac{4}{10}$, $7\frac{3}{10}$, $\frac{750}{100}$, 75, 7.2

3. In the long jump event, Rhonda jumped 1.64 meters. Mary jumped $1\frac{6}{10}$ meters. Kerri jumped $\frac{94}{100}$ meter. Michelle jumped 1.06 meters. Who jumped the farthest?

4. In December, $2\frac{3}{10}$ feet of snow fell. In January, 2.14 feet of snow fell. In February, $2\frac{19}{100}$ feet of snow fell, and in March, $1\frac{1}{10}$ feet of snow fell. During which month did it snow the most? During which month did it snow the least?

Lesson 11: Compare and order mixed numbers in various forms.

©2015 Great Minds. eureka-math.org
G4-M6M7-SE-B4-1.3.1-1.2016

Name _____     Date _____

1.  Plot the following points on the number line using decimal form.

    a.   $0.6$, $\frac{5}{10}$, $0.76$, $\frac{79}{100}$, $0.53$, $\frac{67}{100}$

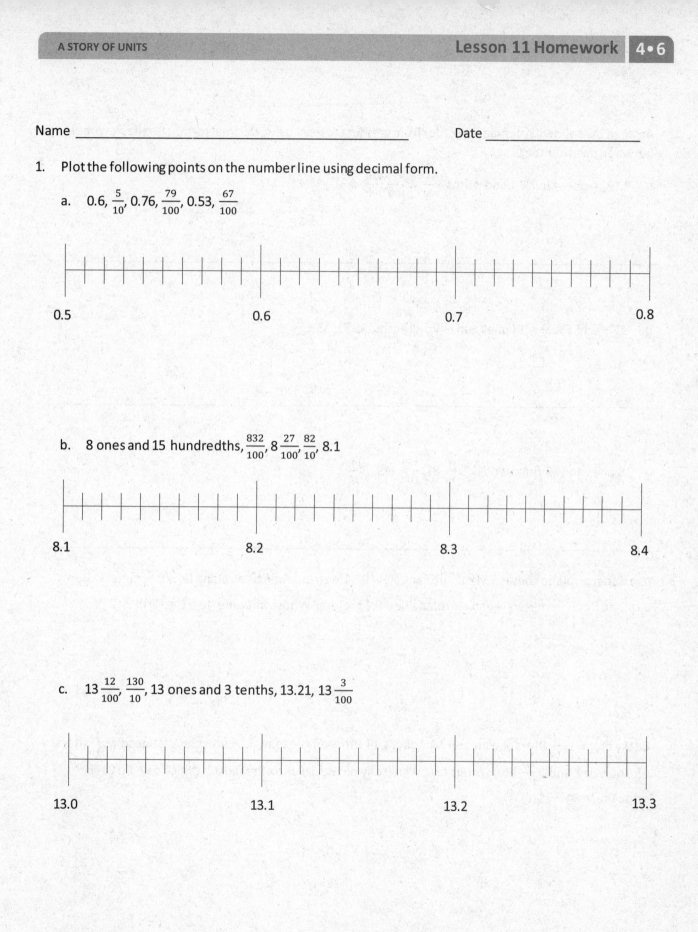

    b.   8 ones and 15 hundredths, $\frac{832}{100}$, $8\frac{27}{100}$, $\frac{82}{10}$, $8.1$

    c.   $13\frac{12}{100}$, $\frac{130}{10}$, 13 ones and 3 tenths, $13.21$, $13\frac{3}{100}$

2. Arrange the following numbers in order from greatest to least using decimal form. Use the > symbol between each number.

   a. 4.03, 4 ones and 33 hundredths, $\frac{34}{100}$, $4\frac{43}{100}$, $\frac{430}{100}$, 4.31

   b. $17\frac{5}{10}$, 17.55, $\frac{157}{10}$, 17 ones and 5 hundredths, 15.71, $15\frac{75}{100}$

   c. 8 ones and 19 hundredths, $9\frac{8}{10}$, 81, $\frac{809}{100}$, 8.9, $8\frac{1}{10}$

3. In a paper airplane contest, Matt's airplane flew 9.14 meters. Jenna's airplane flew $9\frac{4}{10}$ meters. Ben's airplane flew $\frac{904}{100}$ meters. Leah's airplane flew 9.1 meters. Whose airplane flew the farthest?

4. Becky drank $1\frac{41}{100}$ liters of water on Monday, 1.14 liters on Tuesday, 1.04 liters on Wednesday, $\frac{11}{10}$ liters on Thursday, and $1\frac{40}{100}$ liters on Friday. Which day did Becky drink the most? Which day did Becky drink the least?

**Lesson 11:**     Compare and order mixed numbers in various forms.

Name _____ Date _____

1. Complete the number sentence by expressing each part using hundredths. Model using the place value chart, as shown in part (a).

| ones | | tenths | hundredths |
|---|---|---|---|
| | ● | | • • • • • |
| | | | • • • • • |

a. 1 tenth + 5 hundredths = _____ hundredths

| ones | | tenths | hundredths |
|---|---|---|---|
| | ● | | |
| | | | |

b. 2 tenths + 1 hundredth = _____ hundredths

| ones | | tenths | hundredths |
|---|---|---|---|
| | ● | | |
| | | | |

c. 1 tenth + 12 hundredths = _____ hundredths

2. Solve by converting all addends to hundredths before solving.

a. 1 tenth + 3 hundredths = _____ hundredths + 3 hundredths = _____ hundredths

b. 5 tenths + 12 hundredths = _____ hundredths + _____ hundredths = _____ hundredths

c. 7 tenths + 27 hundredths = _____ hundredths + _____ hundredths = _____ hundredths

d. 37 hundredths + 7 tenths = _____ hundredths + _____ hundredths = _____ hundredths

Lesson 12: Apply understanding of fraction equivalence to add tenths and hundredths.

67

©2015 Great Minds. eureka-math.org
G4-M6M7-SE-B4-1.3.1-1.2016

3.  Find the sum.  Convert tenths to hundredths as needed.  Write your answer as a decimal.

    a.  $\frac{2}{10} + \frac{8}{100}$

    b.  $\frac{13}{100} + \frac{4}{10}$

    c.  $\frac{6}{10} + \frac{39}{100}$

    d.  $\frac{70}{100} + \frac{3}{10}$

4.  Solve.  Write your answer as a decimal.

    a.  $\frac{9}{10} + \frac{42}{100}$

    b.  $\frac{70}{100} + \frac{5}{10}$

    c.  $\frac{68}{100} + \frac{8}{10}$

    d.  $\frac{7}{10} + \frac{87}{100}$

5.  Beaker A has $\frac{63}{100}$ liter of iodine.  It is filled the rest of the way with water up to 1 liter.  Beaker B has $\frac{4}{10}$ liter of iodine.  It is filled the rest of the way with water up to 1 liter.  If both beakers are emptied into a large beaker, how much iodine does the large beaker contain?

**Lesson 12:**     Apply understanding of fraction equivalence to add tenths and hundredths.

©2015 Great Minds. eureka-math.org
G4-M6M7-SE-B4-1.3.1 1.2016

Name _____  Date _____

1.  Complete the number sentence by expressing each part using hundredths. Model using the place value chart, as shown in part (a).

| ones | | tenths | hundredths |
|------|--|--------|------------|
| | • | | |

a.  1 tenth + 8 hundredths = _____ hundredths

| ones | | tenths | hundredths |
|------|--|--------|------------|
| | • | | |

b.  2 tenths + 3 hundredths = _____ hundredths

| ones | | tenths | hundredths |
|------|--|--------|------------|
| | • | | |

c.  1 tenth + 14 hundredths = _____ hundredths

2.  Solve by converting all addends to hundredths before solving.

a.  1 tenth + 2 hundredths = _____ hundredths + 2 hundredths = _____ hundredths

b.  4 tenths + 11 hundredths = _____ hundredths + _____ hundredths = _____ hundredths

c.  8 tenths + 25 hundredths = _____ hundredths + _____ hundredths = _____ hundredths

d.  43 hundredths + 6 tenths = _____ hundredths + _____ hundredths = _____ hundredths

Lesson 12:  Apply understanding of fraction equivalence to add tenths and hundredths.

69

©2015 Great Minds. eureka-math.org
G4-M6M7-SE-B4-1.3.1-1.2016

3. Find the sum. Convert tenths to hundredths as needed. Write your answer as a decimal.

a. $\frac{3}{10} + \frac{7}{100}$

b. $\frac{16}{100} + \frac{5}{10}$

c. $\frac{5}{10} + \frac{40}{100}$

d. $\frac{20}{100} + \frac{8}{10}$

4. Solve. Write your answer as a decimal.

a. $\frac{5}{10} + \frac{53}{100}$

b. $\frac{27}{100} + \frac{8}{10}$

c. $\frac{4}{10} + \frac{78}{100}$

d. $\frac{98}{100} + \frac{7}{10}$

5. Cameron measured $\frac{65}{100}$ inch of rainwater on the first day of April. On the second day of April, he measured $\frac{83}{100}$ inch of rainwater. How many total inches of rainwater did Cameron measure on the first two days of April?

**Lesson 12:**    Apply understanding of fraction equivalence to add tenths and hundredths.

©2015 Great Minds. eureka-math.org
G4-M6M7-SE-B4-1.3.1-1.2016

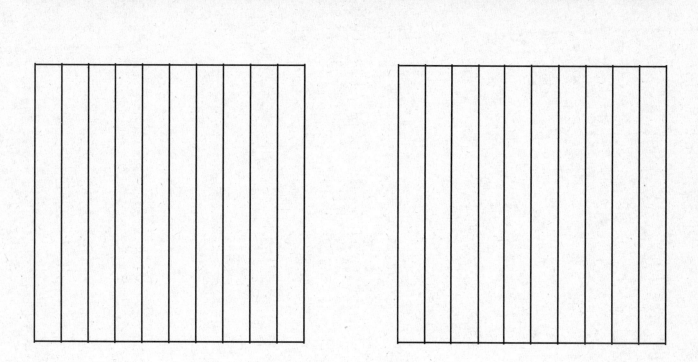

| ones | • | tenths | hundredths |
|------|---|--------|------------|
|      |   |        |            |

area model and place value chart

Lesson 12:    Apply understanding of fraction equivalence to add tenths and
              hundredths.

71

This page intentionally left blank

Name _____ Date _____

1. Solve. Convert tenths to hundredths before finding the sum. Rewrite the complete number sentence in decimal form. Problems 1(a) and 1(b) are partially completed for you.

| | |
|---|---|
| a.  $2\frac{1}{10} + \frac{3}{100} = 2\frac{10}{100} + \frac{3}{100} =$ _____  $2.1 + 0.03 =$ _____ | b.  $2\frac{1}{10} + 5\frac{3}{100} = 2\frac{10}{100} + 5\frac{3}{100} =$ _____ |
| c.  $3\frac{24}{100} + \frac{7}{10}$ | d.  $3\frac{24}{100} + 8\frac{7}{10}$ |

2. Solve. Then, rewrite the complete number sentence in decimal form.

| | |
|---|---|
| a.  $6\frac{9}{10} + 1\frac{10}{100}$ | b.  $9\frac{9}{10} + 2\frac{45}{100}$ |
| c.  $2\frac{4}{10} + 8\frac{90}{100}$ | d.  $6\frac{37}{100} + 7\frac{7}{10}$ |

EUREKA MATH

Lesson 13:    Add decimal numbers by converting to fraction form.

73

3. Solve by rewriting the expression in fraction form. After solving, rewrite the number sentence in decimal form.

| | |
|---|---|
| a.   6.4 + 5.3 | b.   6.62 + 2.98 |
| c.   2.1 + 0.94 | d.   2.1 + 5.94 |
| e.   5.7 + 4.92 | f.   5.68 + 4.9 |
| g.   4.8 + 3.27 | h.   17.6 + 3.59 |

**Lesson 13:**    Add decimal numbers by converting to fraction form.

©2015 Great Minds. eureka-math.org
G4-M6M7-SE-B4-1.3.1-1.2016

Name _____ Date _____

1. Solve. Convert tenths to hundredths before finding the sum. Rewrite the complete number sentence in decimal form. Problems 1(a) and 1(b) are partially completed for you.

| | |
|---|---|
| a. $5\frac{2}{10} + \frac{7}{100} = 5\frac{20}{100} + \frac{7}{100} =$ _____<br><br>$5.2 + 0.07 =$ _____ | b. $5\frac{2}{10} + 3\frac{7}{100} = 8\frac{20}{100} + \frac{7}{100} =$ _____ |
| c. $6\frac{5}{10} + \frac{1}{100}$ | d. $6\frac{5}{10} + 7\frac{1}{100}$ |

2. Solve. Then, rewrite the complete number sentence in decimal form.

| | |
|---|---|
| a. $4\frac{9}{10} + 5\frac{10}{100}$ | b. $8\frac{7}{10} + 2\frac{65}{100}$ |
| c. $7\frac{3}{10} + 6\frac{87}{100}$ | d. $5\frac{48}{100} + 7\frac{8}{10}$ |

**Lesson 13:** Add decimal numbers by converting to fraction form.

75

3.  Solve by rewriting the expression in fraction form.  After solving, rewrite the number sentence in decimal form.

| | |
|---|---|
| a.  $2.1 + 0.87 = 2\frac{1}{10} + \frac{87}{100}$ | b.  $7.2 + 2.67$ |
| c.  $7.3 + 1.8$ | d.  $7.3 + 1.86$ |
| e.  $6.07 + 3.93$ | f.  $6.87 + 3.9$ |
| g.  $8.6 + 4.67$ | h.  $18.62 + 14.7$ |

Lesson 13:    Add decimal numbers by converting to fraction form.

©2015 Great Minds. eureka-math.org
G4-M6M7-SE-B4-1.3.1-1.2016

Name _____ Date _____

1. Barrel A contains 2.7 liters of water. Barrel B contains 3.09 liters of water. Together, how much water do the two barrels contain?

2. Alissa ran a distance of 15.8 kilometers one week and 17.34 kilometers the following week. How far did she run in the two weeks?

Lesson 14: Solve word problems involving the addition of measurements in decimal form.

77

3. An apple orchard sold 140.5 kilograms of apples in the morning and 15.85 kilograms more apples in the afternoon than in the morning. How many total kilograms of apples were sold that day?

4. A team of three ran a relay race. The final runner's time was the fastest, measuring 29.2 seconds. The middle runner's time was 1.89 seconds slower than the final runner's. The starting runner's time was 0.9 seconds slower than the middle runner's. What was the team's total time for the race?

**Lesson 14:**    Solve word problems involving the addition of measurements in decimal form.

©2015 Great Minds. eureka-math.org
G4-M6M7-SE-B4-1.3.1-1.2016

Name _____  Date _____

1. The snowfall in Year 1 was 2.03 meters. The snowfall in Year 2 was 1.6 meters. How many total meters of snow fell in Years 1 and 2?

2. A deli sliced 22.6 kilograms of roast beef one week and 13.54 kilograms the next. How many total kilograms of roast beef did the deli slice in the two weeks?

Lesson 14: Solve word problems involving the addition of measurements in decimal form.

79

©2015 Great Minds. eureka-math.org
G4-M6M7-SE-B4-1.3.1-1.2016

3.  The school cafeteria served 125.6 liters of milk on Monday and 5.34 more liters of milk on Tuesday than on Monday.  How many total liters of milk were served on Monday and Tuesday?

4.  Max, Maria, and Armen were a team in a relay race.  Max ran his part in 17.3 seconds.  Maria was 0.7 seconds slower than Max.  Armen was 1.5 seconds slower than Maria.  What was the total time for the team?

©2015 Great Minds. eureka-math.org
G4-M6M7-SE-B4-1.3.1-1.2016

Name _____     Date _____

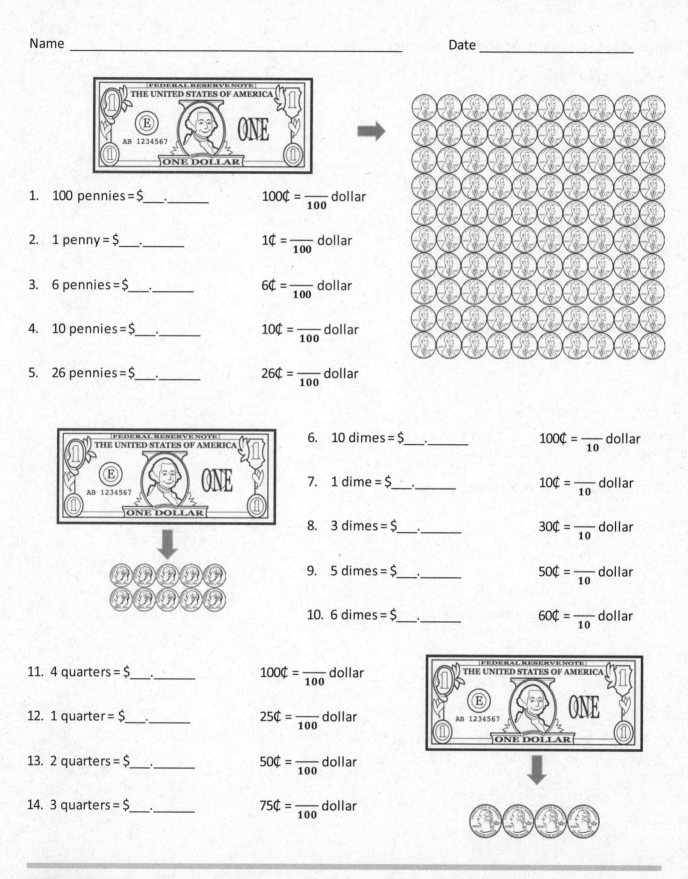

1. 100 pennies = $____._____     100¢ = ——— dollar
                                          100

2. 1 penny = $____._____     1¢ = ——— dollar
                                      100

3. 6 pennies = $____._____     6¢ = ——— dollar
                                      100

4. 10 pennies = $____._____     10¢ = ——— dollar
                                        100

5. 26 pennies = $____._____     26¢ = ——— dollar
                                        100

6. 10 dimes = $____._____     100¢ = ——— dollar
                                       10

7. 1 dime = $____._____     10¢ = ——— dollar
                                     10

8. 3 dimes = $____._____     30¢ = ——— dollar
                                     10

9. 5 dimes = $____._____     50¢ = ——— dollar
                                     10

10. 6 dimes = $____._____     60¢ = ——— dollar
                                      10

11. 4 quarters = $____._____     100¢ = ——— dollar
                                          100

12. 1 quarter = $____._____     25¢ = ——— dollar
                                        100

13. 2 quarters = $____._____     50¢ = ——— dollar
                                         100

14. 3 quarters = $____._____     75¢ = ——— dollar
                                         100

Solve. Give the total amount of money in fraction and decimal form.

15. 3 dimes and 8 pennies

16. 8 dimes and 23 pennies

17. 3 quarters 3 dimes and 5 pennies

18. 236 cents is what fraction of a dollar?

Solve. Express the answer as a decimal.

19. 2 dollars 17 pennies + 4 dollars 2 quarters

20. 3 dollars 8 dimes + 1 dollar 2 quarters 5 pennies

21. 9 dollars 9 dimes + 4 dollars 3 quarters 16 pennies

**Lesson 15:**   Express money amounts given in various forms as decimal numbers.

©2015 Great Minds. eureka-math.org
G4-M6M7-SE-B4-1.3.1-1.2016

Name _____    Date _____

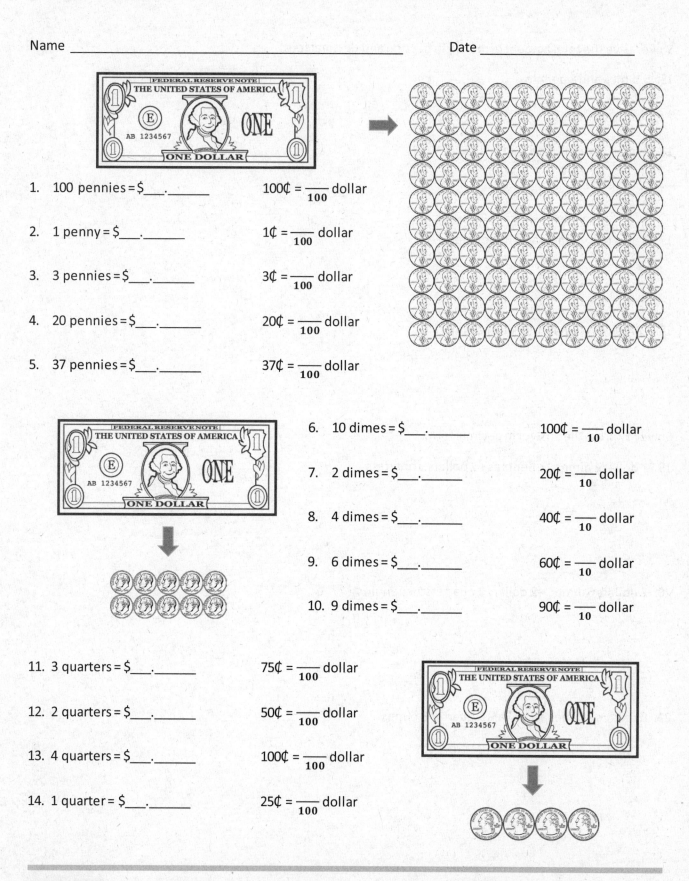

1. 100 pennies = $___._____        100¢ = —— dollar
                                            100

2. 1 penny = $___._____            1¢ = —— dollar
                                          100

3. 3 pennies = $___._____          3¢ = —— dollar
                                          100

4. 20 pennies = $___._____         20¢ = —— dollar
                                           100

5. 37 pennies = $___._____         37¢ = —— dollar
                                           100

6. 10 dimes = $___._____           100¢ = —— dollar
                                            10

7. 2 dimes = $___._____            20¢ = —— dollar
                                           10

8. 4 dimes = $___._____            40¢ = —— dollar
                                           10

9. 6 dimes = $___._____            60¢ = —— dollar
                                           10

10. 9 dimes = $___._____           90¢ = —— dollar
                                           10

11. 3 quarters = $___._____        75¢ = —— dollar
                                           100

12. 2 quarters = $___._____        50¢ = —— dollar
                                           100

13. 4 quarters = $___._____        100¢ = —— dollar
                                            100

14. 1 quarter = $___._____         25¢ = —— dollar
                                           100

Solve. Give the total amount of money in fraction and decimal form.

15. 5 dimes and 8 pennies

16. 3 quarters and 13 pennies

17. 3 quarters 7 dimes and 16 pennies

18. 187 cents is what fraction of a dollar?

Solve. Express the answer in decimal form.

19. 1 dollar 2 dimes 13 pennies + 2 dollars 3 quarters

20. 2 dollars 6 dimes + 2 dollars 2 quarters 16 pennies

21. 8 dollars 8 dimes + 7 dollars 1 quarter 8 dimes

Name _____    Date _____

Use the RDW process to solve.  Write your answer as a decimal.

1. Miguel has 1 dollar bill, 2 dimes, and 7 pennies.  John has 2 dollar bills, 3 quarters, and 9 pennies. How much money do the two boys have in all?

2. Suilin needs 7 dollars 13 cents to buy a book.  In her wallet, she finds 3 dollar bills, 4 dimes, and 14 pennies.  How much more money does Suilin need to buy the book?

3. Vanessa has 6 dimes and 2 pennies.  Joachim has 1 dollar, 3 dimes, and 5 pennies.  Jimmy has 5 dollars and 7 pennies.  They want to put their money together to buy a game that costs $8.00.  Do they have enough money to buy the game?  If not, how much more money do they need?

©2015 Great Minds. eureka-math.org
G4-M6M7-SE-B4-1.3.1-1.2016

4. A pen costs $2.29. A calculator costs 3 times as much as a pen. How much do a pen and a calculator cost together?

5. Krista has 7 dollars and 32 cents. Malory has 2 dollars and 4 cents. How much money does Krista need to give Malory so that each of them has the same amount of money?

Name _____     Date _____

Use the RDW process to solve.  Write your answer as a decimal.

1.   Maria has 2 dollars, 3 dimes, and 4 pennies.  Lisa has 1 dollar and 5 quarters.  How much money do the two girls have in all?

2.   Meiling needs 5 dollars 35 cents to buy a ticket to a show.  In her wallet, she finds 2 dollar bills, 11 dimes, and 5 pennies.  How much more money does Meiling need to buy the ticket?

3.   Joe has 5 dimes and 4 pennies.  Jamal has 2 dollars, 4 dimes, and 5 pennies.  Jimmy has 6 dollars and 4 dimes.  They want to put their money together to buy a book that costs $10.00.  Do they have enough?  If not, how much more do they need?

4.  A package of mechanical pencils costs $4.99.  A package of pens costs twice as much as a package of pencils.  How much do a package of pens and a package of pencils cost together?

5.  Carlos has 8 dollars and 48 cents.  Alissa has 4 dollars and 14 cents.  How much money does Carlos need to give Alissa so that each of them has the same amount of money?

**Lesson 16:**     Solve word problems involving money.

# Eureka Math
## Grade 4
## Module 7

Special thanks go to the Gordon A. Cain Center and to the Department of Mathematics at Louisiana State University for their support in the development of *Eureka Math*.

Printed in the U.S.A.

This book may be purchased from the publisher at eureka-math.org

10  9  8  7  6  5  4  3

ISBN 978-1-63255-306-5

Name _____ Date _____

a.

| Pounds | Ounces |
|--------|--------|
| 1 | |
| 2 | |
| 3 | |
| 4 | |
| 5 | |
| 6 | |
| 7 | |
| 8 | |
| 9 | |
| 10 | |

The rule for converting pounds to ounces is _____.

b.

| Yards | Feet |
|-------|------|
| 1 | |
| 2 | |
| 3 | |
| 4 | |
| 5 | |
| 6 | |
| 7 | |
| 8 | |
| 9 | |
| 10 | |

The rule for converting yards to feet is
_____.

c.

| Feet | Inches |
|------|--------|
| 1 | |
| 2 | |
| 3 | |
| 4 | |
| 5 | |
| 6 | |
| 7 | |
| 8 | |
| 9 | |
| 10 | |

The rule for converting feet to inches is
_____.

EUREKA
MATH™

Lesson 1: Create conversion tables for length, weight, and capacity units using measurement tools, and use the tables to solve problems.

1

©2015 Great Minds. eureka-math.org
G4-M6M7-SE-B4-1.3.1-1.2016

This page intentionally left blank

Name _____ Date _____

Use RDW to solve Problems 1–3.

1. Evan put a 2-pound weight on one side of the scale. How many 1-ounce weights will he need to put on the other side of the scale to make them equal?

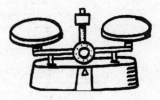

2. Julius put a 3-pound weight on one side of the scale. Abel put 35 1-ounce weights on the other side. How many more 1-ounce weights does Abel need to balance the scale?

3. Mrs. Upton's baby weighs 5 pounds and 4 ounces. How many total ounces does the baby weigh?

4. Complete the following conversion tables, and write the rule under each table.

a.

| Pounds | Ounces |
|--------|--------|
| 1 | |
| 3 | |
| 7 | |
| 10 | |
| 17 | |

The rule for converting pounds to ounces is _____.

Lesson 1: Create conversion tables for length, weight, and capacity units using measurement tools, and use the tables to solve problems.

3

©2015 Great Minds. eureka-math.org
G4-M6M7-SE-B4-1.3.1-1.2016

b.

| Feet | Inches |
|------|--------|
| 1    |        |
| 2    |        |
| 5    |        |
| 10   |        |
| 15   |        |

The rule for converting feet to inches is

_____.

c.

| Yards | Feet |
|-------|------|
| 1     |      |
| 2     |      |
| 4     |      |
| 10    |      |
| 14    |      |

The rule for converting yards to feet is

_____.

5.  Solve.

   a.   3 feet 1 inch = _____ inches

   b.   11 feet 10 inches = _____ inches

   c.   5 yards 1 foot = _____ feet

   d.   12 yards 2 feet = _____ feet

   e.   27 pounds 10 ounces = _____ ounces

   f.   18 yards 9 feet = _____ feet

   g.   14 pounds 5 ounces = _____ ounces

   h.   5 yards 2 feet = _____ inches

6.  Answer *true* or *false* for the following statements.  If the statement is false, change the right side of the comparison to make it true.

   a.   2 kilograms > 2,600 grams        _____

   b.   12 feet < 140 inches             _____

   c.   10 kilometers = 10,000 meters    _____

Lesson 1:        Create conversion tables for length, weight, and capacity units using
                           measurement tools, and use the tables to solve problems.

©2015 Great Minds. eureka-math.org
G4-M6M7-SE-B4-1.3.1-1.2016

Name _____  Date _____

1.  Complete the tables.

a.

| Yards | Feet |
|-------|------|
| 1     |      |
| 2     |      |
| 3     |      |
| 5     |      |
| 10    |      |

b.

| Feet | Inches |
|------|--------|
| 1    |        |
| 2    |        |
| 5    |        |
| 10   |        |
| 15   |        |

c.

| Yards | Inches |
|-------|--------|
| 1     |        |
| 3     |        |
| 6     |        |
| 10    |        |
| 12    |        |

2.  Solve.

a.  2 yards 2 inches = _____ inches

b.  9 yards 10 inches = _____ inches

c.  4 yards 2 feet = _____ feet

d.  13 yards 1 foot = _____ feet

e.  17 feet 2 inches = _____ inches

f.  11 yards 1 foot = _____ feet

g.  15 yards 2 feet = _____ feet

h.  5 yards 2 feet = _____ inches

3.  Ally has a piece of string that is 6 yards 2 feet long.  How many inches of string does she have?

**EUREKA MATH**

**Lesson 1:**  Create conversion tables for length, weight, and capacity units using measurement tools, and use the tables to solve problems.

5

4.  Complete the table.

| Pounds | Ounces |
|--------|--------|
| 1 | |
| 2 | |
| 4 | |
| 10 | |
| 12 | |

5.  Renee's baby sister weighs 7 pounds 2 ounces.  How many ounces does her sister weigh?

6.  Answer *true* or *false* for the following statements.  If the statement is false, change the right side of the comparison to make it true.

a.  4 kilograms < 4,100 grams          _____

b.  10 yards < 360 inches          _____

c.  10 liters = 100,000 milliliters          _____

**Lesson 1:**     Create conversion tables for length, weight, and capacity units using measurement tools, and use the tables to solve problems.

©2015 Great Minds. eureka-math.org
G4-M6M7-SE-B4-1.3.1-1.2016

Name _____    Date _____

a.

| Gallons | Quarts |
|---------|--------|
| 1 | |
| 2 | |
| 3 | |
| 4 | |
| 5 | |
| 6 | |
| 7 | |
| 8 | |
| 9 | |
| 10 | |

The rule for converting gallons to quarts is

_____.

b.

| Quarts | Pints |
|--------|-------|
| 1 | |
| 2 | |
| 3 | |
| 4 | |
| 5 | |
| 6 | |
| 7 | |
| 8 | |
| 9 | |
| 10 | |

The rule for converting quarts to pints is

_____.

c.

| Pints | Cups |
|-------|------|
| 1 | |
| 2 | |
| 3 | |
| 4 | |
| 5 | |
| 6 | |
| 7 | |
| 8 | |
| 9 | |
| 10 | |

The rule for converting pints to cups is _____.

d.    1 gallon = _____ pints

1 quart = _____ cups

1 gallon = _____ cups

Lesson 2:    Create conversion tables for length, weight, and capacity units using measurement tools, and use the tables to solve problems.

7

©2015 Great Minds. eureka-math.org
G4-M6M7-SE-B4-1.3.1-1.2016

This page intentionally left blank

Name _____     Date _____

Use RDW to solve Problems 1–3.

1.  Susie has 3 quarts of milk.  How many pints does she have?

2.  Kristin has 3 gallons 2 quarts of water.  Alana needs the same amount of water but only has 8 quarts.  How many more quarts of water does Alana need?

3.  Leonard bought 4 liters of orange juice.  How many milliliters of juice does he have?

4.  Complete the following conversion tables and write the rule under each table.

a.

| Gallons | Quarts |
|---------|--------|
| 1       |        |
| 3       |        |
| 5       |        |
| 10      |        |
| 13      |        |

The rule for converting gallons to quarts is

_____ .

b.

| Quarts | Pints |
|--------|-------|
| 1      |       |
| 2      |       |
| 6      |       |
| 10     |       |
| 16     |       |

The rule for converting quarts to pints is

_____ .

Lesson 2:    Create conversion tables for length, weight, and capacity units using measurement tools, and use the tables to solve problems.

©2015 Great Minds. eureka-math.org
G4-M6M7-SE-B4-1.3.1-1.2016                                                                              9

5.  Solve.

    a.  8 gallons 2 quarts = _____ quarts

    b.  15 gallons 2 quarts = _____ quarts

    c.  8 quarts 2 pints = _____ pints

    d.  12 quarts 3 pints = _____ cups

    e.  26 gallons 3 quarts = _____ pints

    f.  32 gallons 2 quarts = _____ cups

6.  Answer true or false for the following statements.  If your answer is false, make the statement true.

    a.  1 gallon > 4 quarts        _____

    b.  5 liters = 5,000 milliliters        _____

    c.  15 pints < 1 gallon 1 cup        _____

7.  Russell has 5 liters of a certain medicine.  If it takes 2 milliliters to make 1 dose, how many doses can he make?

8.  Each month, the Moore family drinks 16 gallons of milk and the Siler family goes through 44 quarts of milk.  Which family drinks more milk each month?

9.  Keith's lemonade stand served lemonade in glasses with a capacity of 1 cup.  If he had 9 gallons of lemonade, how many cups could he sell?

Lesson 2:    Create conversion tables for length, weight, and capacity units using measurement tools, and use the tables to solve problems.

©2015 Great Minds. eureka-math.org
G4-M6M7-SE-B4-1.3.1-1.2016

Name _____ Date _____

Use the RDW process to solve Problems 1–3.

1.  Dawn needs to pour 3 gallons of water into her fish tank. She only has a 1-cup measuring cup. How many cups of water should she put in the tank?

2.  Julia has 4 gallons 2 quarts of water. Ally needs the same amount of water but only has 12 quarts. How much more water does Ally need?

3.  Sean drank 2 liters of water today, which was 280 milliliters more than he drank yesterday. How much water did he drink yesterday?

4.  Complete the tables.

    a.

    | Gallons | Quarts |
    |---------|--------|
    | 1       |        |
    | 2       |        |
    | 4       |        |
    | 12      |        |
    | 15      |        |

    b.

    | Quarts | Pints |
    |--------|-------|
    | 1      |       |
    | 2      |       |
    | 6      |       |
    | 10     |       |
    | 16     |       |

**Lesson 2:**  Create conversion tables for length, weight, and capacity units using measurement tools, and use the tables to solve problems.

11

©2015 Great Minds. eureka-math.org
G4-M6M7-SE-B4-1.3.1-1.2016

5.   Solve.

   a.   6 gallons 3 quarts = _____ quarts          b.   12 gallons 2 quarts = _____ quarts

   c.   5 quarts 1 pint = _____ pints             d.   13 quarts 3 pints = _____ cups

   e.   17 gallons 2 quarts = _____ pints          f.   27 gallons 3 quarts = _____ cups

6.   Explain how you solved Problem 5(f).

7.   Answer true or false for the following statements.  If your answer is false, make the statement true by correcting the right side of the comparison.

   a.   2 quarts > 10 pints          _____

   b.   6 liters = 6,000 milliliters          _____

   c.   16 cups < 4 quarts 1 cup          _____

8.   Joey needs to buy 3 quarts of chocolate milk.  The store only sells it in pint containers.  How many pints of chocolate milk should he buy?  Explain how you know.

9.   Granny Smith made punch.  She used 2 pints of ginger ale, 3 pints of fruit punch, and 1 pint of orange juice.  She served the punch in glasses that had a capacity of 1 cup.  How many cups can she fill?

**Lesson 2:**      Create conversion tables for length, weight, and capacity units using measurement tools, and use the tables to solve problems.

©2015 Great Minds. eureka-math.org
G4-M6M7-SE-B4-1.3.1-1.2016

Name _____        Date _____

a.

| Minutes | Seconds |
|---------|---------|
| 1 | |
| 2 | |
| 3 | |
| 4 | |
| 5 | |
| 6 | |
| 7 | |
| 8 | |
| 9 | |
| 10 | |

The rule for converting minutes to seconds is

_____.

b.

| Hours | Minutes |
|-------|---------|
| 1 | |
| 2 | |
| 3 | |
| 4 | |
| 5 | |
| 6 | |
| 7 | |
| 8 | |
| 9 | |
| 10 | |

The rule for converting hours to minutes is

_____.

c.

| Days | Hours |
|------|-------|
| 1 | |
| 2 | |
| 3 | |
| 4 | |
| 5 | |
| 6 | |
| 7 | |
| 8 | |
| 9 | |
| 10 | |

The rule for converting days to hours is

_____.

Lesson 3:    Create conversion tables for units of time, and use the tables to solve
             problems.                                                              13

©2015 Great Minds. eureka-math.org
G4-M6M7-SE-B4-1.3.1-1.2016

This page  intentionally left  blank

Name _____     Date _____

Use RDW to solve Problems 1–2.

1.  Courtney needs to leave the house by 8:00 a.m.  If she wakes up at 6:00 a.m., how many minutes does she have to get ready?  Use the number line to show your work.

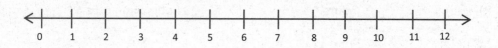

2.  Giuliana's goal was to run a marathon in under 6 hours.  What was her goal in minutes?

3.  Complete the following conversion tables and write the rule under each table.

a.

| Hours | Minutes |
|-------|---------|
| 1     |         |
| 3     |         |
| 6     |         |
| 10    |         |
| 15    |         |

The rule for converting hours to minutes and minutes to seconds is

_____.

b.

| Days | Hours |
|------|-------|
| 1    |       |
| 2    |       |
| 5    |       |
| 7    |       |
| 10   |       |

The rule for converting days to hours is

_____.

Lesson 3:      Create conversion tables for units of time, and use the tables to solve problems.

4.  Solve.

    a.  9 hours 30 minutes = _____ minutes

    b.  7 minutes 45 seconds = _____ seconds

    c.  9 days 20 hours = _____ hours

    d.  22 minutes 27 seconds = _____ seconds

    e.  13 days 19 hours = _____ hours

    f.  23 hours 5 minutes = _____ minutes

5.  Explain how you solved Problem 4(f).

6.  How many seconds are in 14 minutes 43 seconds?

7.  How many hours are there in 4 weeks 3 days?

Lesson 3:  Create conversion tables for units of time, and use the tables to solve problems.

Name _____    Date _____

Use RDW to solve Problems 1–2.

1. Jeffrey practiced his drums from 4:00 p.m. until 7:00 p.m.  How many minutes did he practice?  Use the number line to show your work.

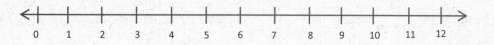

2. Isla used her computer for 5 hours over the weekend.  How many minutes did she spend on the computer?

3. Complete the following conversion tables and write the rule under each table.

a.

| Hours | Minutes |
|-------|---------|
| 1     |         |
| 2     |         |
| 5     |         |
| 9     |         |
| 12    |         |

The rule for converting hours to minutes is

_____.

b.

| Days | Hours |
|------|-------|
| 1    |       |
| 3    |       |
| 6    |       |
| 8    |       |
| 20   |       |

The rule for converting days to hours is

_____.

**Lesson 3:**   Create conversion tables for units of time, and use the tables to solve problems.

17

4.   Solve.

   a.   10 hours 30 minutes = _____ minutes          b.   6 minutes 15 seconds = _____ seconds

   c.   4 days 20 hours = _____ hours                d.   3 minutes 45 seconds = _____ seconds

   e.   23 days 21 hours = _____ hours               f.   17 hours 5 minutes = _____ minutes

5.   Explain how you solved Problem 4(f).

6.   It took a space shuttle 8 minutes 36 seconds to launch and reach outer space.  How many seconds did it take?

7.   Apollo 16's mission lasted just over 1 week 4 days.  How many hours are there in 1 week 4 days?

Name _____     Date _____

Use RDW to solve the following problems.

1.  Beth is allowed 2 hours of TV time each week.  Her sister is allowed 2 times as much.  How many minutes of TV can Beth's sister watch?

2.  Clay weighs 9 times as much as his baby sister.  Clay weighs 63 pounds.  How much does his baby sister weigh in ounces?

3.  Helen has 4 yards of rope.  Daniel has 4 times as much rope as Helen.  How many more feet of rope does Daniel have compared to Helen?

Lesson 4:    Solve multiplicative comparison word problems using measurement conversion tables.

19

©2015 Great Minds. eureka-math.org
G4-M6M7-SE-B4-1.3.1-1.2016

4.  A dishwasher uses 11 liters of water for each cycle.  A washing machine uses 5 times as much water as a dishwasher uses for each load.  Combined, how many milliliters of water are used for 1 cycle of each machine?

5.  Joyce bought 2 pounds of apples.  She bought 3 times as many pounds of potatoes as pounds of apples.  The melons she bought were 10 ounces lighter than the total weight of the potatoes.  How many ounces did the melons weigh?

Name _____     Date _____

Use RDW to solve the following problems.

1.  Sandy took the train to New York City.  The trip took 3 hours.  Jackie took the bus, which took twice as long.  How many minutes did Jackie's trip take?

2.  Coleton's puppy weighed 3 pounds 8 ounces at birth.  The vet weighed the puppy again at 6 months, and the puppy weighed 7 pounds.  How many ounces did the puppy gain?

3.  Jessie bought a 2-liter bottle of juice.  Her sister drank 650 milliliters.  How many milliliters were left in the bottle?

©2015 Great Minds. eureka-math.org
G4-M6M7-SE-B4-1.3.1-1.2016

4. Hudson has a chain that is 1 yard in length. Myah's chain is 3 times as long. How many feet of chain do they have in all?

5. A box weighs 8 ounces. A shipment of boxes weighs 7 pounds. How many boxes are in the shipment?

6. Tracy's rain barrel has a capacity of 27 quarts of water. Beth's rain barrel has a capacity of twice the amount of water as Tracy's rain barrel. Trevor's rain barrel can hold 9 quarts of water less than Beth's barrel.

   a. What is the capacity of Trevor's rain barrel?

   b. If Tracy, Beth, and Trevor's rain barrels were filled to capacity, and they poured all of the water into a 30-gallon bucket, would there be enough room? Explain.

**Lesson 4:** Solve multiplicative comparison word problems using measurement conversion tables.

©2015 Great Minds. eureka-math.org
G4-M6M7-SE-B4-1.3.1-1.2016

Name _____   Date _____

1.  a.  Label the rest of the tape diagram below.  Solve for the unknown.

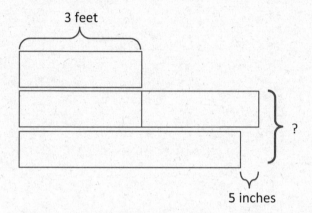

3 feet

?

5 inches

    b.  Write a problem of your own that could be solved using the diagram above.

2.  Create a problem of your own using the diagram below, and solve for the unknown.

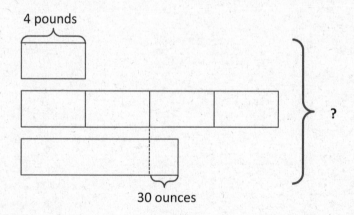

4 pounds

?

30 ounces

This page  intentionally left  blank

Name _____    Date _____

Draw a tape diagram to solve the following problems.

1.  Timmy drank 2 quarts of water yesterday.  He drank twice as much water today as he drank yesterday.
    How many cups of water did Timmy drink in the two days?

2.  Lisa recorded a 2-hour television show.  When she watched it, she skipped the commercials.  It took her
    84 minutes to watch the show.  How many minutes did she save by skipping the commercials?

3.  Jason bought 2 pounds of cashews.  Sarah ate 9 ounces.  David ate 2 ounces more than Sarah.  How many
    ounces were left in Jason's bag of cashews?

4.  a.  Label the rest of the tape diagram below.  Solve for the unknown.

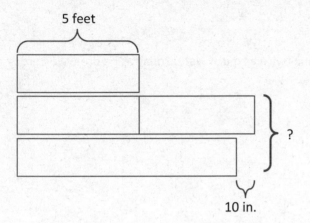

5 feet

?

10 in.

    b.  Write a problem of your own that could be solved using the diagram above.

5.  Create a problem of your own using the diagram below, and solve for the unknown.

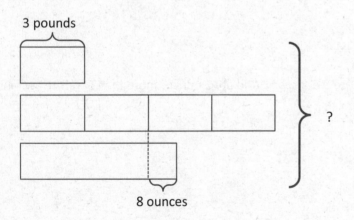

3 pounds

?

8 ounces

EUREKA
MATH™

| Classmate: | | Problem Number: | |
|---|---|---|---|
| Strategies my classmate used: | | | |
| Things my classmate did well: | | | |
| Suggestions for improvement: | | | |
| Changes I would make to my work based on my classmate's work: | | | |

| Classmate: | | Problem Number: | |
|---|---|---|---|
| Strategies my classmate used: | | | |
| Things my classmate did well: | | | |
| Suggestions for improvement: | | | |
| Changes I would make to my work based on my classmate's work: | | | |

peer share and critique form

©2015 Great Minds. eureka-math.org
G4-M6M7-SE-B4-1.3.1-1.2016

This page intentionally left blank

Name _____     Date _____

1.  Determine the following sums and differences.  Show your work.

    a.    3 qt + 1 qt = _____ gal                b.  2 gal 1 qt + 3 qt = _____ gal

    c.    1 gal – 1 qt = _____ qt                d.  5 gal – 1 qt = _____ gal _____ qt

    e.    2 c + 2 c = _____ qt                  f.  1 qt 1 pt + 3 pt = _____ qt

    g.    2 qt – 3 pt = _____ pt                h.  5 qt – 3 c _____ qt _____ c

2.  Find the following sums and differences.  Show your work.

    a.    6 gal 3 qt + 3 qt = _____ gal _____ qt     b.  10 gal 3 qt + 3 gal 3 qt = _____ gal _____ qt

    c.    9 gal 1 pt – 2 pt = _____ gal _____ pt      d.  7 gal 1 pt – 2 gal 7 pt = _____ gal _____ pt

    e.    16 qt 2 c + 4 c = _____ qt _____ c        f.  6 gal 5 pt + 3 gal 3 pt = _____ gal _____ pt

©2015 Great Minds. eureka-math.org
G4-M6M7-SE-B4-1.3.1-1.2016

3.   The capacity of a pitcher is 3 quarts.  Right now, it contains 1 quart 3 cups of liquid.  How much more liquid can the pitcher hold?

4.   Dorothy follows the recipe in the table to make her grandma's cherry lemonade.

   a.   How much lemonade does the recipe make?

| Cherry Lemonade | |
| --- | --- |
| Ingredient | Amount |
| Lemon Juice | 5 pints |
| Sugar Syrup | 2 cups |
| Water | 1 gallon 1 quart |
| Cherry Juice | 3 quarts |

   b.   How many more cups of water could Dorothy add to the recipe to make an exact number of gallons of lemonade?

©2015 Great Minds. eureka-math.org
G4-M6M7-SE-B4-1.3.1-1.2016

Name _____     Date _____

1. Determine the following sums and differences. Show your work.

   a. 5 qt + 3 qt = _____ gal

   b. 1 gal 2 qt + 2 qt = _____ gal

   c. 1 gal – 3 qt = _____ qt

   d. 3 gal – 2 qt = _____ gal _____ qt

   e. 1 c + 3 c = _____ qt

   f. 2 qt 3 c + 5 c = _____ qt

   g. 1 qt – 1 pt = _____ pt

   h. 6 qt – 5 pt = _____ qt _____ pt

2. Find the following sums and differences. Show your work.

   a. 4 gal 2 qt + 3 qt = _____ gal _____ qt

   b. 12 gal 2 qt + 5 gal 3 qt = _____ gal _____ qt

   c. 7 gal 2 pt – 3 pt = _____ gal _____ pt

   d. 11 gal 3 pt – 4 gal 6 pt = _____ gal _____ pt

   e. 12 qt 5 c + 6 c = _____ qt _____ c

   f. 8 gal 6 pt + 5 gal 4 pt = _____ gal _____ pt

©2015 Great Minds. eureka-math.org
G4-M6M7-SE-B4-1.3.1-1.2016

3. The capacity of a bucket is 5 gallons. Right now, it contains 3 gallons 2 quarts of liquid. How much more liquid can the bucket hold?

4. Grace and Joyce follow the recipe in the table to make a homemade bubble solution.

   a. How much solution does the recipe make?

| Homemade Bubble Solution | |
|---|---|
| Ingredient | Amount |
| Water | 2 gallons 3 pints |
| Dish Soap | 2 quarts 1 cup |
| Corn Syrup | 2 cups |

   b. How many more cups of solution would they need to fill a 4-gallon container?

Name _____    Date _____

1. Determine the following sums and differences. Show your work.

   a.  1 ft + 2 ft = _____ yd               b.  3 yd 1 ft + 2 ft = _____ yd

   c.  1 yd − 1 ft = _____ ft               d.  8 yd − 1 ft = _____ yd _____ ft

   e.  3 in + 9 in = _____ ft               f.  6 in + 9 in = _____ ft _____ in

   g.  1 ft − 8 in = _____ in               h.  5 ft − 8 in = _____ ft _____ in

2. Find the following sums and differences. Show your work.

   a.  5 yd 2 ft + 2 ft = _____ yd _____ ft     b.  7 yd 2 ft + 2 yd 2 ft = _____ yd _____ ft

   c.  4 yd 1 ft − 2 ft = _____ yd _____ ft     d.  6 yd 1 ft − 2 yd 2 ft = _____ yd _____ ft

   e.  6 ft 9 in + 4 in = _____ ft _____ in     f.  4 ft 4 in + 3 ft 11 in = _____ ft _____ in

   g.  34 ft 4 in − 8 in = _____ ft _____ in    h.  7 ft 1 in − 5 ft 10 in = _____ ft _____ in

©2015 Great Minds. eureka-math.org
G4-M6M7-SE-B4-1.3.1-1.2016

3.  Matthew is 6 feet 2 inches tall.  His little cousin Emma is 3 feet 6 inches tall.  How much taller is Matthew than Emma?

4.  In gym class, Jared climbed 10 feet 4 inches up a rope.  Then, he continued to climb up another 3 feet 9 inches.  How high did Jared climb?

5.  A quadrilateral has a perimeter of 18 feet 2 inches.  The sum of three of the sides is 12 feet 4 inches.

    a.  What is the length of the fourth side?

    b.  An equilateral triangle has a side length equal to the fourth side of the quadrilateral.  What is the perimeter of the triangle?

Name _____     Date _____

1.  Determine the following sums and differences.  Show your work.

    a.   2 yd 2 ft + 1 ft = _____ yd

    b.   2 yd – 1 ft = _____ yd _____ ft

    b.   2 ft + 2 ft = _____ yd _____ ft

    d.   5 yd – 1 ft = _____ yd _____ ft

    e.   7 in + 5 in = _____ ft

    f.   7 in + 7 in = _____ ft _____ in

    g.   1 ft – 2 in = _____ in

    h.   2 ft – 6 in = _____ ft _____ in

2.  Find the following sums and differences.  Show your work.

    a.   4 yd 2 ft + 2 ft = _____ yd _____ ft

    b.   6 yd 2 ft + 1 yd 1 ft = _____ yd _____ ft

    c.   5 yd 1 ft – 2 ft = _____ yd _____ ft

    d.   7 yd 1 ft – 5 yd 2 ft = _____ yd _____ ft

    e.   7 ft 8 in + 5 in = _____ ft _____ in

    f.   6 ft 5 in + 5 ft 9 in = _____ ft _____ in

    g.   32 ft 3 in – 7 in = _____ ft _____ in

    h.   8 ft 2 in – 3 ft 11 in = _____ ft _____ in

3.  Laurie bought 9 feet 5 inches of blue ribbon.  She also bought 6 feet 4 inches of green ribbon.  How much ribbon did she buy altogether?

4.  The length of the room is 11 feet 6 inches.  The width of the room is 2 feet 9 inches shorter than the length.  What is the width of the room?

5.  Tim's bedroom is 12 feet 6 inches wide.  The perimeter of the rectangular-shaped bedroom is 50 feet.

    a.  What is the length of Tim's bedroom?

    b.  How much longer is the length of Tim's room than the width?

©2015 Great Minds. eureka-math.org
G4-M6M7-SE-B4-1.3.1-1.2016

Name _____     Date _____

1.  Determine the following sums and differences.  Show your work.

    a.   7 oz + 9 oz = _____ lb                     b.   1 lb 5 oz + 11 oz = _____ lb

    c.   1 lb – 13 oz = _____ oz                    d.   12 lb – 4 oz = _____ lb _____ oz

    e.   3 lb 9 oz + 9 oz = _____ lb _____ oz     f.   30 lb 9 oz + 9 lb 9 oz _____ lb _____ oz

    g.   25 lb 2 oz – 14 oz = _____ lb _____ oz   h.   125 lb 2 oz – 12 lb 3 oz = _____ lb _____ oz

2.  The total weight of Sarah and Amanda's full backpacks is 27 pounds.  Sarah's backpack weighs 15 pounds 9 ounces.  How much does Amanda's backpack weigh?

©2015 Great Minds. eureka-math.org
G4-M6M7-SE-B4-1.3.1-1.2016

3.  In Emma's supply box, a pencil weighs 3 ounces.  Her scissors weigh 3 ounces more than the pencil, and a bottle of glue weighs three times as much as the scissors.  How much does the bottle of glue weigh in pounds and ounces?

4.  Use the information in the chart about Jodi's school supplies to answer the following questions:

    a.  On Mondays, Jodi packs only her laptop and supply case into her backpack.  How much does her full backpack weigh?

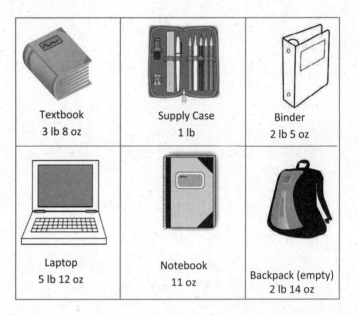

| Textbook 3 lb 8 oz | Supply Case 1 lb | Binder 2 lb 5 oz |
| Laptop 5 lb 12 oz | Notebook 11 oz | Backpack (empty) 2 lb 14 oz |

    b.  On Tuesdays, Jodi brings her laptop, supply case, two notebooks, and two textbooks in her backpack.  On Fridays, Jodi only packs her binder and supply case.  How much less does Jodi's full backpack weigh on Friday than it does on Tuesday?

EUREKA MATH

Name _____     Date _____

1.  Determine the following sums and differences.  Show your work.

    a.   11 oz + 5 oz = _____ lb

    b.   1 lb 7 oz + 9 oz = _____ lb

    c.   1 lb – 11 oz = _____ oz

    d.   12 lb – 8 oz = _____ lb _____ oz

    e.   5 lb 8 oz + 9 oz = _____ lb _____ oz

    f.   21 lb 8 oz + 6 lb 9 oz = _____ lb _____ oz

    g.   23 lb 1 oz – 15 oz = _____ lb _____ oz

    h.   89 lb 2 oz – 16 lb 4 oz = _____ lb _____ oz

2.  When David took his dog, Rocky, to the vet in December, Rocky weighed 29 pounds 9 ounces.  When he took Rocky back to the vet in March, Rocky weighed 34 pounds 4 ounces.  How much weight did Rocky gain?

3.  Bianca had 6 identical jars of bubble bath.  She put them all in a bag that weighed 2 ounces.  The total weight of the bag filled with the six jars was 1 pound 4 ounces.  How much did each jar weigh?

©2015 Great Minds. eureka-math.org
G4-M6M7-SE-B4-1.3.1-1.2016

4. Use the information in the chart about Melissa's school supplies to answer the following questions:

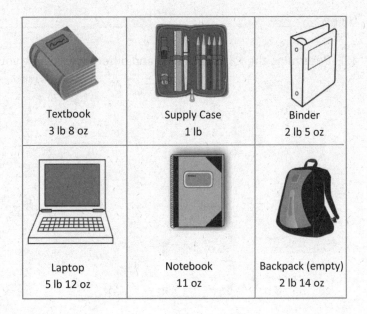

| | | |
|---|---|---|
| Textbook 3 lb 8 oz | Supply Case 1 lb | Binder 2 lb 5 oz |
| Laptop 5 lb 12 oz | Notebook 11 oz | Backpack (empty) 2 lb 14 oz |

a. On Wednesdays, Melissa packs only two notebooks and a binder into her backpack. How much does her full backpack weigh on Wednesdays?

b. On Thursdays, Melissa puts her laptop, supply case, two textbooks, and a notebook in her backpack. How much does her full backpack weigh on Thursdays?

c. How much more does the backpack weigh with 3 textbooks and a notebook than it does with just 1 textbook and the supply case?

Lesson 8: Solve problems involving mixed units of weight.

EUREKA MATH

©2015 Great Minds. eureka-math.org
G4-M6M7-SE-B4-1.3.1-1.2016

Name _____     Date _____

1. Determine the following sums and differences. Show your work.

   a. 23 min + 37 min = _____ hr

   b. 1 hr 11 min + 49 min = _____ hr

   c. 1 hr – 12 min = _____ min

   d. 4 hr – 12 min = _____ hr _____ min

   e. 22 sec + 38 sec = _____ min

   f. 3 min – 45 sec = _____ min _____ sec

2. Find the following sums and differences. Show your work.

   a. 3 hr 45 min + 25 min = _____ hr _____ min

   b. 2 hr 45 min + 6 hr 25 min = _____ hr _____ min

   c. 3 hr 7 min – 42 min = _____ hr _____ min

   d. 5 hr 7 min – 2 hr 13 min = _____ hr _____ min

   e. 5 min 40 sec + 27 sec = _____ min _____ sec

   f. 22 min 48 sec – 5 min 58 sec = ____ min ____ sec

©2015 Great Minds. eureka-math.org
G4-M6M7-SE-B4-1.3.1-1.2016

3. At the cup-stacking competition, the first place finishing time was 1 minute 52 seconds. That was 31 seconds faster than the second place finisher. What was the second place time?

4. Jackeline and Raychel have 5 hours to watch three movies that last 1 hour 22 minutes, 2 hours 12 minutes, and 1 hour 57 minutes, respectively.

   a. Do the girls have enough time to watch all three movies? Explain why or why not.

   b. If Jackeline and Raychel decide to watch only the two longest movies and take a 30-minute break in between, how much of their 5 hours will they have left over?

Lesson 9:     Solve problems involving mixed units of time.

©2015 Great Minds. eureka-math.org
G4-M6M7 SE B4-1.3.1-1.2016

Name _____ Date _____

1. Determine the following sums and differences. Show your work.

   a. 41 min + 19 min = _____ hr

   b. 2 hr 21 min + 39 min = _____ hr

   c. 1 hr – 33 min = _____ min

   d. 3 hr – 33 min = _____ hr _____ min

   e. 31 sec + 29 sec = _____ min

   f. 5 min – 15 sec = _____ min _____ sec

2. Find the following sums and differences. Show your work.

   a. 5 hr 30 min + 35 min = _____ hr _____ min

   b. 3 hr 15 min + 5 hr 55 min = _____ hr _____ min

   c. 4 hr 4 min – 38 min = _____ hr _____ min

   d. 7 hr 3 min – 4 hr 25 min = _____ hr _____ min

   e. 3 min 20 sec + 49 sec = _____ min _____ sec

   f. 22 min 37 sec – 5 min 58 sec = _____ min _____ sec

©2015 Great Minds. eureka-math.org
G4-M6M7-SE-B4-1.3.1-1.2016

3. It took 5 minutes 34 seconds for Melissa's oven to preheat to 350 degrees. That was 27 seconds slower than it took Ryan's oven to preheat to the same temperature. How long did it take Ryan's oven to preheat?

4. Joanna read three books. Her goal was to finish all three books in a total of 7 hours. She completed them, respectively, in 2 hours 37 minutes, 3 hours 9 minutes, and 1 hour 51 minutes.

    a. Did Joanna meet her goal? Write a statement to explain why or why not.

    b. Joanna completed the two shortest books in one evening. How long did she spend reading that evening? How long, with her goal in mind, did that leave her to read the third book?

Name _____    Date _____

Use RDW to solve the following problems.

1. Paula's time swimming in the Ironman Triathlon was 1 hour 25 minutes. Her time biking was 5 hours longer than her swimming time. She ran for 4 hours 50 minutes. How long did it take her to complete all three parts of the race?

2. Nolan put 7 gallons 3 quarts of gas into his car on Monday and twice as much on Saturday. What was the total amount of gas put into the car on both days?

3.  One pumpkin weighs 7 pounds 12 ounces.  A second pumpkin weighs 10 pounds 4 ounces.  A third pumpkin weighs 2 pounds 9 ounces more than the second pumpkin.  What is the total weight of all three pumpkins?

4.  Mr. Lane is 6 feet 4 inches tall.  His daughter, Mary, is 3 feet 8 inches shorter than her father.  His son is 9 inches taller than Mary.  How many inches taller is Mr. Lane than his son?

**Lesson 10:**     Solve multi-step measurement word problems.

©2015 Great Minds. eureka-math.org
G4-M6M7-SE-B4-1.3.1-1.2016

Name _____    Date _____

Use RDW to solve the following problems.

1.  On Saturday, Jeff used 2 quarts 1 cup of water from a full gallon to replace some water that leaked from his fish tank.  On Sunday, he used 3 pints of water from the same gallon.  How much water was left in the gallon after Sunday?

2.  To make punch, Julia poured 1 quart 3 cups of ginger ale into a bowl and then added twice as much fruit juice.  How much punch did she make in all?

3.  Patti went swimming for 1 hour 15 minutes on Monday.  On Tuesday, she swam twice as long as she swam on Monday.  On Wednesday, she swam 50 minutes less than the time she swam on Tuesday.  How much time did she spend swimming during that three-day period?

4.  Myah is 4 feet 2 inches tall.  Her sister, Ally, is 10 inches taller.  Their little brother is half as tall as Ally. How tall is their little brother in feet and inches?

5.  Rick and Laurie have three dogs.  Diesel weighs 89 pounds 12 ounces.  Ebony weighs 33 pounds 14 ounces less than Diesel.  Luna is the smallest at 10 pounds 2 ounces.  What is the combined weight of the three dogs in pounds and ounces?

**Lesson 10:**     Solve multi-step measurement word problems.

©2015 Great Minds. eureka-math.org
G4-M6M7-SE-B4-1.3.1-1.2016

Name _____     Date _____

Use RDW to solve the following problems.

1.  Lauren ran a marathon and finished 1 hour 15 minutes after Amy, who had a time of 2 hours 20 minutes. Cassie finished 35 minutes after Lauren.  How long did it take Cassie to run the marathon?

2.  Chef Joe has 8 lb 4 oz of ground beef in his freezer.  This is $\frac{1}{3}$ of the amount needed to make the number of burgers he planned for a party.  If he uses 4 oz of beef for each burger, how many burgers is he planning to make?

3.  Sarah read for 1 hour 17 minutes each day for 6 days.  If she took 3 minutes to read each page, how many pages did she read in 6 days?

4.  Grades 3, 4, and 5 have their annual field day together.  Each grade level is given 16 gallons of water. If there are a total of 350 students, will there be enough water for each student to have 2 cups?

Name _____  Date _____

Use RDW to solve the following problems.

1. Ashley ran a marathon and finished 1 hour 40 minutes after P.J., who had a time of 2 hours 15 minutes. Kerry finished 12 minutes before Ashley.  How long did it take Kerry to run the marathon?

2. Mr. Foote's deck is 12 ft 6 in wide.  Its length is twice the width plus 3 more inches.  How long is the deck?

3. Mrs. Lorentz bought 12 pounds 8 ounces of sugar.  This is $\frac{1}{4}$ of the sugar she will use to make sugar cookies in her bakery this week.  If she uses 10 ounces of sugar for each batch of sugar cookies, how many batches of sugar cookies will she make in a week?

4. Beth Ann practiced piano for 1 hour 5 minutes each day for 1 week. She had 5 songs to practice and spent the same amount of time practicing each song. How long did she practice each song during the week?

5. The concession stand has 18 gallons of punch. If there are a total of 240 students who want to purchase 1 cup of punch each, will there be enough punch for everyone?

**Lesson 11:**      Solve multi-step measurement word problems.

©2015 Great Minds. eureka-math.org
G4-M6M7-SE-B4-1.3.1-1.2016

Name _____   Date _____

1.  Draw a tape diagram to show 1 yard divided into 3 equal parts.

    a.  $\frac{1}{3}$ yd = _____ ft

    b.  $\frac{2}{3}$ yd = _____ ft

    c.  $\frac{3}{3}$ yd = _____ ft

2.  Draw a tape diagram to show $2\frac{2}{3}$ yards = 8 feet.

3.  Draw a tape diagram to show $\frac{3}{4}$ gallon = 3 quarts.

4.  Draw a tape diagram to show $3\frac{3}{4}$ gallons = 15 quarts.

5.  Solve the problems using whatever tool works best for you.

    a.  $\frac{1}{12}$ ft = _____ in

    b.  $\frac{\phantom{0}}{12}$ ft = $\frac{1}{2}$ ft = _____ in

    c.  $\frac{\phantom{0}}{12}$ ft = $\frac{1}{4}$ ft = _____ in

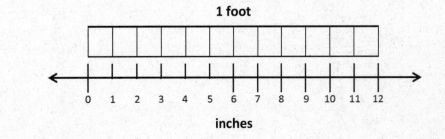

Lesson 12:   Use measurement tools to convert mixed number measurements to smaller units.

53

d. $\dfrac{}{12}$ ft = $\dfrac{3}{4}$ ft = _____ in

e. $\dfrac{}{12}$ ft = $\dfrac{1}{3}$ ft = _____ in

f. $\dfrac{}{12}$ ft = $\dfrac{2}{3}$ ft = _____ in

6. Solve.

| | |
|---|---|
| a. $1\dfrac{1}{3}$ yd = _____ ft | b. $4\dfrac{2}{3}$ yd = _____ ft |
| c. $2\dfrac{1}{2}$ gal = _____ qt | d. $7\dfrac{3}{4}$ gal = _____ qt |
| e. $1\dfrac{1}{2}$ ft = _____ in | f. $6\dfrac{1}{2}$ ft = _____ in |
| g. $1\dfrac{1}{4}$ ft = _____ in | h. $6\dfrac{1}{4}$ ft = _____ in |

**Lesson 12:** Use measurement tools to convert mixed number measurements to smaller units.

EUREKA
MATH

Name _____    Date _____

1. Draw a tape diagram to show $1\frac{1}{3}$ yards = 4 feet.

2. Draw a tape diagram to show $\frac{1}{2}$ gallon = 2 quarts.

3. Draw a tape diagram to show $1\frac{3}{4}$ gallons = 7 quarts.

4. Solve the problems using whatever tool works best for you.

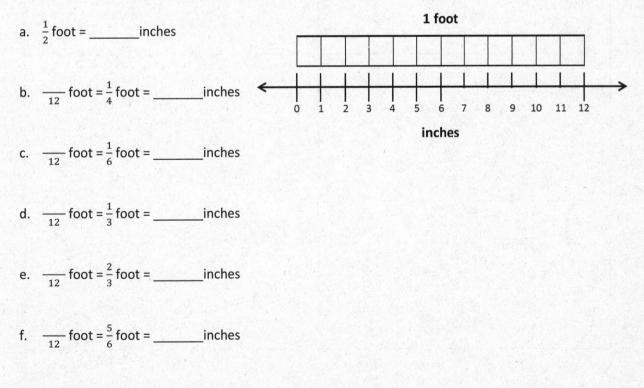

   a. $\frac{1}{2}$ foot = _____ inches

   b. $\frac{}{12}$ foot = $\frac{1}{4}$ foot = _____ inches

   c. $\frac{}{12}$ foot = $\frac{1}{6}$ foot = _____ inches

   d. $\frac{}{12}$ foot = $\frac{1}{3}$ foot = _____ inches

   e. $\frac{}{12}$ foot = $\frac{2}{3}$ foot = _____ inches

   f. $\frac{}{12}$ foot = $\frac{5}{6}$ foot = _____ inches

Lesson 12:   Use measurement tools to convert mixed number measurements to smaller units.

55

©2015 Great Minds. eureka-math.org
G4-M6M7-SE-B4-1.3.1-1.2016

5. Solve.

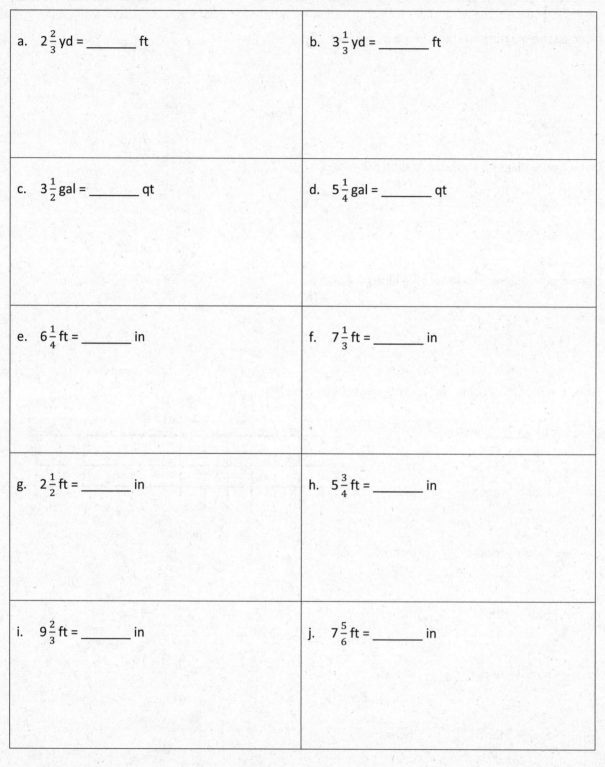

| | |
|---|---|
| a. $2\frac{2}{3}$ yd = _____ ft | b. $3\frac{1}{3}$ yd = _____ ft |
| c. $3\frac{1}{2}$ gal = _____ qt | d. $5\frac{1}{4}$ gal = _____ qt |
| e. $6\frac{1}{4}$ ft = _____ in | f. $7\frac{1}{3}$ ft = _____ in |
| g. $2\frac{1}{2}$ ft = _____ in | h. $5\frac{3}{4}$ ft = _____ in |
| i. $9\frac{2}{3}$ ft = _____ in | j. $7\frac{5}{6}$ ft = _____ in |

**Lesson 12:** Use measurement tools to convert mixed number measurements to smaller units.

©2015 Great Minds. eureka-math.org
G4-M6M7-SE-B4-1.3.1-1.2016

EUREKA MATH™

Name _____    Date _____

1.  Solve.

    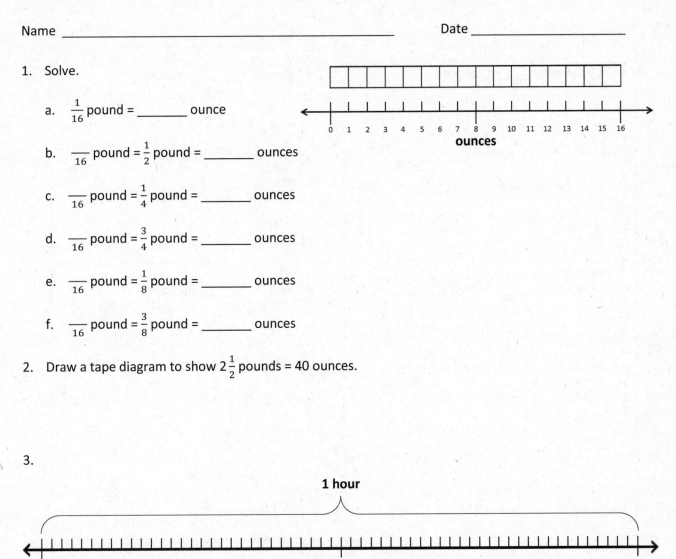

    a.  $\frac{1}{16}$ pound = _____ ounce

    b.  $\frac{}{16}$ pound = $\frac{1}{2}$ pound = _____ ounces

    c.  $\frac{}{16}$ pound = $\frac{1}{4}$ pound = _____ ounces

    d.  $\frac{}{16}$ pound = $\frac{3}{4}$ pound = _____ ounces

    e.  $\frac{}{16}$ pound = $\frac{1}{8}$ pound = _____ ounces

    f.  $\frac{}{16}$ pound = $\frac{3}{8}$ pound = _____ ounces

2.  Draw a tape diagram to show $2\frac{1}{2}$ pounds = 40 ounces.

3.

    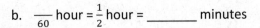

    a.  $\frac{1}{60}$ hour = _____ minute

    b.  $\frac{}{60}$ hour = $\frac{1}{2}$ hour = _____ minutes

    c.  $\frac{}{60}$ hour = $\frac{1}{4}$ hour = _____ minutes

4.  Draw a tape diagram to show that $1\frac{1}{2}$ hours = 90 minutes.

Lesson 13:   Use measurement tools to convert mixed number measurements to smaller units.

57

©2015 Great Minds. eureka-math.org
G4-M6M7-SE-B4-1.3.1-1.2016

5. Solve.

| | |
|---|---|
| a.  $1\frac{1}{8}$ pounds = _____ ounces | b.  $3\frac{3}{8}$ pounds = _____ ounces |
| c.  $5\frac{3}{4}$ lb = _____ oz | d.  $5\frac{1}{2}$ lb = _____ oz |
| e.  $1\frac{1}{4}$ hours = _____ minutes | f.  $3\frac{1}{2}$ hours = _____ minutes |
| g.  $2\frac{1}{4}$ hr = _____ min | h.  $5\frac{1}{2}$ hr = _____ min |
| i.  $3\frac{1}{3}$ yards = _____ feet | j.  $7\frac{2}{3}$ yd = _____ ft |
| k.  $4\frac{1}{2}$ gallons = _____ quarts | l.  $6\frac{3}{4}$ gal = _____ qt |
| m.  $5\frac{3}{4}$ feet = _____ inches | n.  $8\frac{1}{3}$ ft = _____ in |

**Lesson 13:**    Use measurement tools to convert mixed number measurements to smaller units.

©2015 Great Minds. eureka-math.org
G4-M6M7-SE-B4-1.3.1-1.2016

Name _____     Date _____

1.  Solve.

    a.   $\frac{1}{16}$ pound = _____ ounce

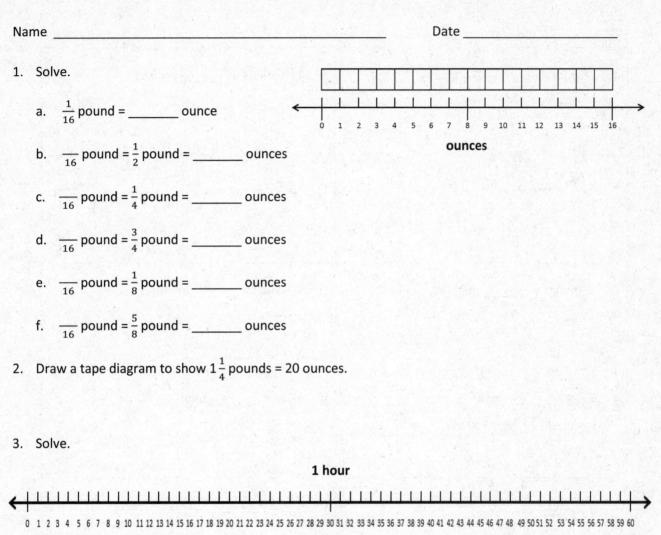

    b.   $\frac{}{16}$ pound = $\frac{1}{2}$ pound = _____ ounces

    c.   $\frac{}{16}$ pound = $\frac{1}{4}$ pound = _____ ounces

    d.   $\frac{}{16}$ pound = $\frac{3}{4}$ pound = _____ ounces

    e.   $\frac{}{16}$ pound = $\frac{1}{8}$ pound = _____ ounces

    f.   $\frac{}{16}$ pound = $\frac{5}{8}$ pound = _____ ounces

2.  Draw a tape diagram to show $1\frac{1}{4}$ pounds = 20 ounces.

3.  Solve.

**1 hour**

0 1 2 3 4 5 6 7 8 9 10 11 12 13 14 15 16 17 18 19 20 21 22 23 24 25 26 27 28 29 30 31 32 33 34 35 36 37 38 39 40 41 42 43 44 45 46 47 48 49 50 51 52 53 54 55 56 57 58 59 60

**minutes**

    a.   $\frac{1}{60}$ hour = _____ minute

    b.   $\frac{}{60}$ hour = $\frac{1}{2}$ hour = _____ minutes

    c.   $\frac{}{60}$ hour = $\frac{1}{4}$ hour = _____ minutes

    d.   $\frac{}{60}$ hour = $\frac{1}{3}$ hour = _____ minutes

4.  Draw a tape diagram to show that $2\frac{1}{4}$ hours = 135 minutes.

Lesson 13:      Use measurement tools to convert mixed number measurements to           **59**
                smaller units.

5. Solve.

| | |
|---|---|
| a. $2\frac{1}{4}$ pounds = _____ ounces | b. $4\frac{7}{8}$ pounds = _____ ounces |
| c. $6\frac{3}{4}$ lb = _____ oz | d. $4\frac{1}{8}$ lb = _____ oz |
| e. $1\frac{3}{4}$ hours = _____ minutes | f. $4\frac{1}{2}$ hours = _____ minutes |
| g. $3\frac{3}{4}$ hr = _____ min | h. $5\frac{1}{3}$ hr = _____ min |
| i. $4\frac{2}{3}$ yards = _____ feet | j. $6\frac{1}{3}$ yd = _____ ft |
| k. $4\frac{1}{4}$ gallons = _____ quarts | l. $2\frac{3}{4}$ gal = _____ qt |
| m. $6\frac{1}{4}$ feet = _____ inches | n. $9\frac{5}{6}$ ft = _____ in |

Lesson 13: Use measurement tools to convert mixed number measurements to smaller units.

EUREKA MATH

Name _____   Date _____

Use RDW to solve the following problems.

1. A cartoon lasts $\frac{1}{2}$ hour. A movie is 6 times as long as the cartoon. How many minutes does it take to watch both the cartoon and the movie?

2. A large bench is $7\frac{1}{6}$ feet long. It is 17 inches longer than a shorter bench. How many inches long is the shorter bench?

3. The first container holds 4 gallons 2 quarts of juice. The second container can hold $1\frac{3}{4}$ gallons more than the first container. Altogether, how much juice can the two containers hold?

Lesson 14:    Solve multi-step word problems involving converting mixed number
             measurements to a single unit.

©2015 Great Minds. eureka-math.org
G4-M6M7-SE-B4-1.3.1-1.2016                                                          61

4. A girl's height is $3\frac{1}{3}$ feet. A giraffe's height is 3 times that of the girl's. How many inches taller is the giraffe than the girl?

5. Five ounces of pretzels are put into each bag. How many bags can be made from $22\frac{3}{4}$ pounds of pretzels?

6. Twenty servings of pancakes require 15 ounces of pancake mix.

   a. How much pancake mix is needed for 120 servings?

   b. Extension: The mix is bought in $2\frac{1}{2}$-pound bags. How many bags will be needed to make 120 servings?

**Lesson 14:**   Solve multi-step word problems involving converting mixed number measurements to a single unit.

Name _____   Date _____

Use RDW to solve the following problems.

1.  Molly baked a pie for 1 hour and 45 minutes.  Then, she baked banana bread for 35 minutes less than the pie.  How many minutes did it take to bake the pie and the bread?

2.  A slide on the playground is $12\frac{1}{2}$ feet long.  It is 3 feet 7 inches longer than the small slide.  How long is the small slide?

3.  The fish tank holds 8 gallons 2 quarts of water.  Jeffrey poured $1\frac{3}{4}$ gallons into the empty tank.  How much more water does he still need to pour into the tank to fill it?

Lesson 14:    Solve multi-step word problems involving converting mixed number
              measurements to a single unit.

©2015 Great Minds. eureka-math.org
G4-M6M7-SE-B4-1.3.1-1.2016

63

4. The candy shop puts 10 ounces of gummy bears in each box. How many boxes do they need to fill if there are $21\frac{1}{4}$ pounds of gummy bears?

5. Mom can make 10 brownies from a 12-ounce package.

   a. How many ounces of brownie mix would be needed to make 50 brownies?

   b. Extension: The brownie mix is also sold in $1\frac{1}{2}$-pound bags. How many bags would be needed to make 120 brownies?

Name _____     Date _____

1. Emma's rectangular bedroom is 11 ft long and 12 ft wide with an attached closet that is 4 ft by 5 ft. How many square feet of carpet does Emma need to cover both the bedroom and closet?

2. To save money, Emma is no longer going to carpet her closet. In addition, she wants one 3 ft by 6 ft corner of her bedroom to be wood floor. How many square feet of carpet will she need for the bedroom now?

EUREKA
MATH™

Lesson 15:      Create and determine the area of composite figures.

65

©2015 Great Minds. eureka-math.org
G4-M6M7-SE-B4-1.3.1-1.2016

3. Find the area of the figure pictured to the right.

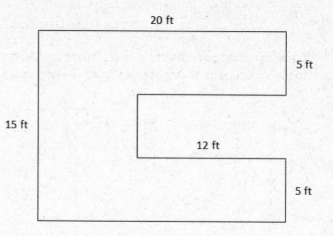

4. Label the sides of the figure below with measurements that make sense. Find the area of the figure.

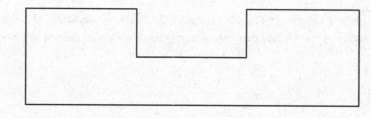

**Lesson 15:**    Create and determine the area of composite figures.

©2015 Great Minds. eureka-math.org
G4-M6M7-SE-B4-1.3.1-1.2016

5.  Peterkin Park has a square fountain with a walkway around it.  The fountain measures 12 feet on each side.  The walkway is $3\frac{1}{2}$ feet wide.  Find the area of the walkway.

6.  If 1 bag of gravel covers 9 square feet, how many bags of gravel will be needed to cover the entire walkway around the fountain in Peterkin Park?

This page intentionally left blank

Name _____     Date _____

For homework, complete the top portion of each page.  This will become an answer key for you to refer to when completing the bottom portion as a mini-personal white board activity during the summer.

Find the area of the figure that is shaded.

1.

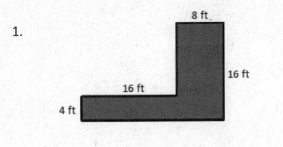

2.

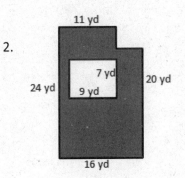

- - - - - - - - - - - - - - - - - - - - - - - - - - - - - - - - - - - - - - - - - - - - - - - - - - - -

Find the area of the figure that is shaded.

1.

2.

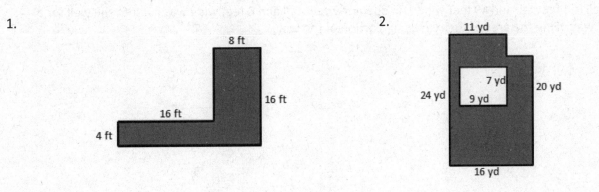

Challenge:  Replace the given dimensions with different measurements, and solve again.

3. A wall is 8 feet tall and 19 feet wide. An opening 7 feet tall and 8 feet wide was cut into the wall for a doorway. Find the area of the remaining portion of the wall.

- - - - - - - - - - - - - - - - - - - - - - - - - - - - - - - - - - - - - - - - - - - - - - - - - - - -

3. A wall is 8 feet tall and 19 feet wide. An opening 7 feet tall and 8 feet wide was cut into the wall for a doorway. Find the area of the remaining portion of the wall.

©2015 Great Minds. eureka-math.org
G4-M6M7-SE-B4-1.3.1-1.2016

Name _____     Date _____

Work with your partner to create each floor plan on a separate piece of paper, as described below.

You should use a protractor and a ruler to create each floor plan and be sure each rectangle you create has two sets of parallel lines and four right angles.

Be sure to label each part of your model with the correct measurement.

1. The bedroom in Samantha's dollhouse is a rectangle 26 centimeters long and 15 centimeters wide. It has a rectangular bed that is 9 centimeters long and 6 centimeters wide. The two dressers in the room are each 2 centimeters wide. One measures 7 centimeters long, and the other measures 4 centimeters long. Create a floor plan of the bedroom containing the bed and dressers. Find the area of the open floor space in the bedroom after the furniture is in place.

2. A model of a rectangular pool is 15 centimeters long and 10 centimeters wide. The walkway around the pool is 5 centimeters wider than the pool on each of the four sides. In one section of the walkway, there is a flowerbed that is 3 centimeters by 5 centimeters. Create a diagram of the pool area with the surrounding walkway and flowerbed. Find the area of the open walkway around the pool.

This page  intentionally left  blank

Name _____     Date _____

For homework, complete the top portion of each page.  This will become an answer key for you to refer to when completing the bottom portion as a mini-personal white board activity during the summer.

Use a ruler and protractor to create and shade a figure according to the directions.  Then, find the area of the unshaded part of the figure.

1.  Draw a rectangle that is 18 cm long and 6 cm wide.  Inside the rectangle, draw a smaller rectangle that is 8 cm long and 4 cm wide.  Inside the smaller rectangle, draw a square that has a side length of 3 cm.  Shade in the smaller rectangle, but leave the square unshaded.  Find the area of the unshaded space.

- - - - - - - - - - - - - - - - - - - - - - - - - - - - - - - - - - - - - - - - - - - - - - -

1.  Draw a rectangle that is 18 cm long and 6 cm wide.  Inside the rectangle, draw a smaller rectangle that is 8 cm long and 4 cm wide.  Inside the smaller rectangle, draw a square that has a side length of 3 cm.  Shade in the smaller rectangle, but leave the square unshaded.  Find the area of the unshaded space.

©2015 Great Minds. eureka-math.org
G4-M6M7-SE-B4-1.3.1-1.2016

2.  Emanuel's science project display board is 42 inches long and 48 inches wide.  He put a 6-inch border around the edge inside the board and placed a title in the center of the board that is 22 inches long and 6 inches wide.  How many square inches of open space does Emanuel have left on his board?

- - - - - - - - - - - - - - - - - - - - - - - - - - - - - - - - - - - - - - - - - - - - - - - - - - - -

2.  Emanuel's science project display board is 42 inches long and 48 inches wide.  He put a 6-inch border around the edge inside the board and placed a title in the center of the board that is 22 inches long and 6 inches wide.  How many square inches of open space does Emanuel have left on his board?
    Challenge:  Replace the given dimensions with different measurements, and solve again.

©2015 Great Minds. eureka-math.org
G4-M6M7-SE-B4-1.3.1-1.2016

Name _____    Date _____

1.  Decimal Fraction Review: Plot and label each point on the number line below, and complete the chart. Only solve the portion above the dotted line.

| Point | Unit Form | Decimal Form | Mixed Number (ones and fraction form) | How much more to get to the next whole number? |
|-------|-----------|--------------|---------------------------------------|------------------------------------------------|
| A | 2 ones and 9 tenths | | | |
| B | | 4.4 | $4\frac{4}{10}$ | |
| C | | | | $\frac{2}{10}$ or 0.2 |

- - - - - - - - - - - - - - - - - - - - - - - - - - - - - - - - - - - - - - - - -

1.  Complete the chart.  Create your own problem for B, and plot the point.

| Point | Unit Form | Decimal Form | Mixed Number (ones and fraction form) | How much more to get to the next whole number? |
|-------|-----------|--------------|---------------------------------------|------------------------------------------------|
| A | 2 ones and 9 tenths | | | |
| B | | | | |

EUREKA MATH™

**Lesson 17:**    Practice and solidify Grade 4 fluency.

**75**

©2015 Great Minds. eureka-math.org
G4-M6M7-SE-B4-1.3.1-1.2016

2.  Complete the chart.  The first one has been done for you.  Only solve the top portion above the dotted line.

| Decimal | Mixed Number | Tenths | Hundredths |
|---|---|---|---|
| 3.2 | $3\frac{2}{10}$ | 32 tenths   or   $\frac{32}{10}$ | 320 hundredths   or   $\frac{320}{100}$ |
| 8.6 | | | |
| 11.7 | | | |
| 4.8 | | | |

- - - - - - - - - - - - - - - - - - - - - - - - - - - - - - - - - - - - - - - - - - -

2.  Complete the chart.  Create your own problem in the last row.

| Decimal | Mixed Number | Tenths | Hundredths |
|---|---|---|---|
| 3.2 | | | |
| 8.6 | | | |
| 11.7 | | | |
| | | | |

**Lesson 17:**    Practice and solidify Grade 4 fluency.

©2015 Great Minds. eureka-math.org
G4-M6M7-SE-B4-1.3.1-1.2016

This page intentionally left blank

This page intentionally left blank

This page intentionally left blank

This page intentionally left blank

This page intentionally left blank

This page  intentionally left  blank

This page intentionally left blank

This page intentionally left blank